W9-CZT-812

St. Louis Community College

Library

5801 Wilson Avenue
St. Louis, Missouri 63110

Spacefarers of the '80s and '90s
The Next Thousand People in Space

SPACEFARERS OF THE '80s AND '90s

The Next Thousand People in Space

ALCESTIS R. OBERG

New York Columbia University Press *1985*

Library of Congress Cataloging in Publication Data

Oberg, Alcestis R., 1949 –
Spacefarers of the '80s and '90s.

Bibliography: p.
Includes Index.
1. Astronautics. 2. Space colonies. I. Title.
TL790.026 1985   629.45   84-21514
ISBN 0-231-05906-X (alk. paper)

COLUMBIA UNIVERSITY PRESS
NEW YORK   GUILFORD, SURREY

DEDICATION

To Jim
A dreamer who drives a rusty car
and has unlimited faith in the future;
and
To my family and friends,
who have unlimited faith in me.

Contents

Figures 1–30 appear as a group after chapter 3.
Figures 31–58 appear as a group after chapter 11.

Preface

My interest in space began in 1979, on a miserable January day in Houston. There were thunder, lightning, heavy rain, cold and fog — simultaneously — a meteorological phenomenon possible only in Houston. At the time, I was editor of the *L-5 News,* a job I got on the strength of my knowledge of English grammar and my willingness to work for slave wages. I had slogged to the Shamrock Hilton that day to cover a space presentation at a large science convention only because I could not convince someone else to do so.

Nearly a decade before this, I had married James Oberg — before he became a renowned space expert. I'm afraid some future biographer will make much of how we watched the first moon landing together, a month before our wedding, how we were struggling writers for years, both driven by a love of space. It's all baloney. Although we did watch the moon landing together, we didn't struggle much; neither of us had planned on becoming a journalist, and our attraction to one another had nothing to do with space. For many years, whenever space happened to come up in conversation, which wasn't often, I'm afraid I can be accused of listening with the benign boredom that the wives of brain surgeons show when they listen to details of operations and wives of accountants display when they hear the latest skirmishes with the IRS. Space did not claim me until January 1979.

That year was a bad one for space. The conference papers ranged optimistically over future space medicine, future selection of space travelers, future space business opportunities, and so on — all subjects on which I would write later. But there was a strong, unmistakable undercurrent of pessimism because of the consistent

antagonism of the Carter administration to the space program —
the rumor that Carter planned to dissolve NASA as an agency and
make it a minor branch of the Department of Transportation, the
whispered threat that the Office of Management and Budget wanted
to scrap the shuttle program altogether, and so on. The year 1979
was the midpoint in a long void in our space expansion — between
Skylab, which was falling toward Earth, and the Shuttle, which had
not yet left it.

In the private conversations in the restaurant later, the enthusiasts
talked of space. I listened. I touched the tangible, wonderful future
with their minds and wanted to protect it. I had breathed the word
Space for the first time. Like Saul in the Bible, I had met my calling
unexpectedly, on a road going someplace else.

In the months and years that followed, it became more difficult
to turn back to my earlier pursuits and career aims. I continued to
write about space. Each story was an adventure, an expansion, and
each interview was a journey of exploration. That is still true today.

Hundreds of articles later, I have come to believe that this is the
greatest time in the history of the world to be alive. All of us have
been touched by space in one way or another whether we are
conscious of it or not. Our children's lives — indeed, the whole
history of mankind — will be changed by it in the decades to come.
Going to the moon was the greatest endeavor our species has ever
attempted. The seeding of the universe with human intelligence
will be the next.

As the trailblazing exploits of the first 100 people in space passes
into the quiet effort of the first thousand and the first ten thousand,
the real work of civilizing the solar system will begin. The firsts that
are reserved for the shuttle era — the use of the first space tethers,
the first piece of space construction, the first light sail, the first space
mirror, all the firsts of the new technologies to be ushered in by the
new space transportation system — will pass without much fanfare
in the media.

But the second wave of spacefarers — those of the 1980s and
1990s — will be the ultimate civilizers of the space frontier. They
will shape and be shaped by the space environment — its awesome
perspective, its special terrors, its disorienting rhythms, its unutter-
able grace. They will not all be astronauts. Many will be workers

from industry, scientists from universities, writers from the media, and artists, authors, musicians, and other representatives from the fine arts. If we are very lucky, we may find the common man or woman, the spine and sinew of the country, headed toward space. Like the early Puritans who came to wrest a life from the New World, the next thousand people in space will begin to develop the frontier's potential, to establish its industries, and to write its intellectual history. Although there may be some heroes and possibly some geniuses among them, the spacefarers of the 1980s and 1990s will undoubtedly be a faceless, somewhat anonymous group whose major contribution will lie in "being there," seeding near-Earth orbit, the Moon and Mars with space villages—with humanity's presence.

In stages, the next thousand people in space will transform space travel from something extraordinary to something altogether routine; then they will help others make the leap into the future by proving the viability of nascent space industries; finally, they will begin to mold the future by creating a basis for future ventures, the most obvious of which are moon-mining and further interplanetary travel. Along with that, it is their destiny to affirm human virtue in the face of a hostile environment, to show the stuff of which human beings are made. It is for them all—engineers, scientists, architects, explorers, poets, and philosophers — to write the future history of the world, to become the ancestors and founders of the remarkable beings who will be our shining descendants, as different from us as we are from the people of the Middle Ages: the terraformers, the star travelers, the builders and reapers of the universe.

This book is about them.

Acknowledgments

I wish to thank many people whose aid was invaluable in the preparation of this book.

At the Johnson Space Center, John Lawrence, Terri White, and Billie Deeson were extremely helpful in arranging interviews with astronauts or in getting special permission to get into various restricted facilities.

I also thank Mike Gentry and Lisa Vasquez for rounding up those great NASA photos for me.

At Marshall Spaceflight Center, David Drachlis was especially helpful in getting interviews with people involved in various Spacelab flights.

At Air Force Space Division in Los Angeles, Ellen Rose and Lt. Col. Geoff Baker were absolutely terrific in getting permission for me to interview the MSE program manager. I thank the program manager too for taking the time to answer my long list of questions in detail.

I was grateful to NASA Administrator James Beggs for our long and far-ranging discussion on current and future NASA projects, and for his continuing effort to open the space program up to people from many different walks of life, from many different nations of the world.

NASA Headquarters workers William O'Donnell and Cindy Cline gave me invaluable insight into the future operations of NASA.

I thank my old friend Dr. Daniel Woodard for perusing the chapters on medicine and psychology for accuracy.

I owe my deepest debt of gratitude to my editor at Columbia University Press, Craig Waff. His persistent and enthusiastic efforts to acquire this book did more to organize and consolidate my thinking about it than any other factor. His patient and understanding handling of various ruffles that arose in the course of writing and editing the book was greatly appreciated. He has my vote for the first diplomat/troubleshooter in space.

Last but not least, I thank my husband, Jim, who supported, encouraged, sympathized with, and cheered me through the vissicitudes of this first book as he has through many endeavors in our life together.

Spacefarers of the '80s and '90s
The Next Thousand People in Space

1

The Space Experience

A space sunrise is bold, beautiful, and brief. It is not the slow lightening of the horizon but a crescent sword of color brandished between the perpetual stygian darkness of space and the ebony blackness of the sleeping Earth. "You see eight different bands of color come and go," said Joe Allen, "from a brilliant red to the brightest and deepest blue. No sunrise or sunset is ever the same. I mean, they're not like rainbows, which have the same color combinations no matter where you are on Earth. The colors change and the width of the bands is different every time. And you see 16 sunrises and 16 sunsets every day you're in space" (Allen and O'Toole 1983:63).

Time is not measured by the movement of the sun and stars in space. Human beings are set free of celestial events — sunrise, sunset, moonrise, moonset, the position of Venus, the progression of the constellations — as the spacecraft orbits the Earth every 90 minutes, traveling approximately 17,000 miles an hour. When a wake-up call is sent up to the shuttle from Mission Control at 5 A.M. Houston time, the scene outside the spacecraft window may look like blazing noon. "It's been an interesting morning to wake up," said Jack Lousma on STS-3. "It was beautiful and sunny when we pulled up the window shade and looked out—and halfway through the pass, the sun went down and now it's black as night."

Set free of their fundamental Earthborne sense of time, space-crews structure their day according to computerized digital clocks and a timeline, a preordained, carefully orchestrated, often rehearsed succession of tasks.

In the early shuttle program, place was as important as time. The crews used computer-displayed orbit maps to plot their location over the ground in order to anticipate upcoming communication passes over exotic-sounding tracking stations: Yarragadee and Or-roral Valley in Australia, Dakar and Botswana in Africa, Santiago in Chile, Buckhorn and Goldstone in California. The tenuous communication between Earth and spacecraft was usually brief — only a few minutes over some tracking sites — and often garbled as a result of interference. Instructions were always prioritized on the ground before being radioed to the spacecraft through the CAPCOM (capsule communicator), a fellow astronaut in Mission Control who sat to the right of the flight director and filtered all information to the astronauts. The absence of communication was tense mostly during times of danger in the flight — during crucial rocket burns, spacewalks, and fiery reentry. Now the Tracking Data Relay Satellites (TDRS) make it possible for Earth and shuttle to talk to one another during nearly 85 percent of each orbit rather than only 15 percent, as with the old tracking stations.

Launch is an experience that sends even the most unflappable astronaut's pulse shooting up. Seven million pounds of thrust ignites suddenly from below. The solid rockets crackle and pop loudly and the shuttle vibrates and shudders, slowly escaping the intangible fist of gravity. "Your whole soul knows when the solids light," said Joe Allen. "Your body knows you're right in the middle of a sound chamber. Your whole self is shaking and you know you're on the front end of the world's most powerful afterburner, going straight up" (Allen and O'Toole 1983:114). As the rocket ascends, the sky turns from blue to black. The astronauts are pushed back in their seats as the spacecraft accelerates rapidly to the point where their limbs feel twice as heavy as normal and they have to think about breathing.

When the slender solid rockets have finally spent their fury some 15 miles high, they drop off, marking their separation by a very bright flash around all the forward windshields, a sudden blizzard of ice, and a cessation of the terrifying shaking and rattling of the spacecraft. "Looking out through the side window, right at solid rocket booster separation," said Joe Allen, "it looked like it began to snow, and it stayed with us — snow going in all directions, not just streaming back along the vehicle" (air to ground STS-5).

A night launch is slightly different. "It's like being in the middle of a lightning ball," recalled Dan Brandenstein of STS-8. "There's no need for any lights in the cabin at all while the SRBs (solid rocket boosters) are on. It's brighter than daylight in the cabin."

The three main engines, which carry the astronauts the remaining distance into orbit and are fueled by a large external tank of liquid hydrogen, are much quieter and smoother and whir like "a big electric motor," according to Sally Ride of STS-7. That tank is jettisoned too, destined to fall into the Indian Ocean until plans mature someday to carry it all the way into orbit for use in space construction. There is no physiological sensation of separation from the large external tank at all. The only indication is that the main engine lights go out, and the RCS (Reaction Control System) jets begin to fire for attitude control.

The astronauts are not always aware of the exact moment they cross the threshold between Earth's gravity and space's weightlessness. For instance, while still strapped in his seat during the launch of STS-5, Bill Lenoir took a pen out of his pocket and let it go. It floated, assuring his senses that he was finally spaceborne.

Once in orbit, all kinds of debris may float freely both inside and outside the spacecraft. On the first flight of the *Challenger*, screws, bolts, washers, and even a button from some workman's coat floated into the cabin, to be sucked up eventually by the cabin fan filters. A stowaway Florida fruit fly, which apparently entered the cabin before the flight, flew around crazily, while the crew stalked it, hoping it wouldn't get into the mung bean experiment onboard.

Outside the spacecraft an icy Christmas tree might form on the vertical fins. "There's a significant build-up of ice sticking up an ice fence, must be on the engine bell," reported Story Musgrave during STS-6. Piled up like a snowdrift, the ice slowly became desiccated and disappeared after a few days. Also a few thermal tiles along the top of the OMS (Orbital Maneuvering System) pods were shaken loose by the tremendous force of the launch during STS-1, giving rise to concern that other, more crucial tiles had also been shaken loose, a worry that turned out to be totally unjustified. Still, something shakes loose on just about every flight. Joe Allen reported that on STS-5, "chunks of material coming off the OMS pods...look like pieces of paper. They're quite thin and they're about the size of a flake of paper up to and including an 8½ x 11-inch sheet of paper. Very thin." On the same flight, Bill Lenoir reported: "On daylight side passes, looking aft, we see little rings, and little bits of ice, apparently coming off the main engines. We'll see a strip slightly curve, a strip maybe a couple feet long, come off and spin away very slowly." Ice that breaks off "winks because it's rotating in the sunlight as it spins away, like a star pulsing," said Story Musgrave of STS-6.

Once in orbit, the noise level drops dramatically to that "of an airliner" said Don Peterson of STS-6, although the middeck (the downstairs of the vehicle) is noisier than the flight deck where the pilots sit. It is impossible for an astronaut on the middeck to be heard on the flight deck even when shouting, so crewmembers talk over wireless intercoms. The astronauts cannot perceive speed in a spacecraft. There is no air to whistle past the windows, no motion outside that would convey a sense of forward momentum other than continents passing below them. In a press conference, Joe Allen and Norm Thagard described space flying as a kind of quiet coasting, as in a "gondola under a hot-air balloon racing across the sky."

Most important, the space travelers must learn a completely different kind of locomotion from the start. All it takes to get across the Orbiter middeck is a gentle push with the fingertips of one hand. A gentle push with one's feet is plenty to propel a person from the middeck to the flight deck above. Astronauts accustomed to fighting Earth's gravity usually push off with their feet or hands too hard, which results in their banging into everything. However, most adapt quickly.

Because there is no up and down in space, astronauts must mentally orient themselves to get their bearings. Wrote Joe Allen: "You don't know if the spacecraft is right side up or upside down. You know your direction, but that's never related to which way your nose is pointed. You can be flying tail first, wing first, or belly first and you know you're traveling east, but you have to do a deliberate mental calculation that if that's east and the Earth is down there, then over there must be north."(Allen and O'Toole 1983:63).

Story Musgrave and Don Peterson on STS-6 had no disorientation problems either in the spacecraft or on the spacewalk because "we had a moving frame of reference." According to Peterson: "I decided wherever my head was pointed was up. I'd look to see where I was going and just go — and not worry about up and down."

Part of the first day in orbit is spent going through innumerable vehicle checkouts to ensure that the spacecraft is sound. The payload bay doors are opened to get rid of excess heat, to expose experiments on the space pallet, and to prepare to launch satellites. The business of the shuttle revolves around satellites — launching, repairing, and retrieving them. Satellite launching is a simple procedure. The astronauts signal the satellite's protective shroud to peel back, exposing it to space. The satellite then is spun like a top until it is going fast enough to release it safely from the payload bay. The Orbiter then carefully backs away from the satellite. When the shuttle is far away, the satellite rockets fire to carry it into a higher orbit where it parks and starts performing its function. Retrieving a satellite is somewhat trickier, involving careful positioning of the shuttle's long crane-like arm (remote manipulator system or RMS), grappling the satellite, and returning it to the shuttle bay for immediate repair or for return to Earth.

Launching a satellite from the shuttle payload bay differs remarkably from Earth simulations because of the strange environment. Computer dials must replace hearing and feeling: the shuttle itself must become an extension of the spaceperson's body. Joe Allen described it: "When you hear those satellites spin on the ground, and if you're in the same room, it sounds as if a subway train is about to run you over. In space there's nothing to carry sound. In space we looked at that big thing in the payload bay, seven tons of it rotating...and you couldn't hear a whisper. You couldn't even feel a vibration. We knew those satellites were spinning only because our eyes told us they were going around and our computer said they were spining at exactly 49.9 revolutions a minute" (Allen & O'Toole 1983:62).

Although the shuttle was economically justified as a cargo carrier and was designed for that purpose, limited scientific activity can go on, and many experiments look toward the future potential of space. On the middeck, crystal growth and pharmaceutical experiments abound, demonstrating the huge riches these industries may derive from the space environment. Solar flare studies render some benefit to astronomy, although the shuttle's attitude control jets fire far too often to make the craft an ideal observatory. Life science experiments can be a mixed blessing. Attempts to grow pine trees, mung beans, and sponges involve very little of an astronaut's time—usually a simple matter of turning a light switch on or off. Others require observation. On STS-3, an experiment testing how flying insects fare in zero gravity provoked Jack Lousma into saying: "They sure do make noise; it's hard to sleep at night with them buzzing away." Nonetheless, Lousma's observations were interesting: "The moths are very lively, the bees have all gotten stationary, and the flies took to walking. Bees got smart fast. They decided there wasn't any use flapping their wings and going out of control, so they just float and wiggle their legs. As far as the moths go, they like to stand on the sides of the Plexiglas cage too, but lot more of them seem to fly, and some seem to have adapted to flying from one place to another."

Astronauts have a horror of ants, rodents, and other such creatures that can do real damage to the electronics of the shuttle if they should get loose.

The astronauts themselves are guinea pigs for ongoing experiments on human adaptation to zero gravity. On STS-5, Joe Allen and Bill Lenoir did an acceleration detection sensitivity test in which they measured the levels at which the body detects linear and angular acceleration — that is, how the eye, brain, and inner ear perceive motion in space. The test involved placing an astronaut in a harness, suspending him by 4 springs from the ceiling of the middeck, making him wear an eyeshield and earplugs, and spinning him until he could perceive motion. Five electrodes on the face monitored involuntary responses.

On Spacelab-1, payload specialist Byron Lichtenberg and mission specialist Owen Garriot tested the reactions of the central nervous system to space by hanging suspended midair while their reflexes were systematically stimulated by a series of tiny electrical shocks on the leg, called "the drop and shock test."

Spacewalks using rocket-powered, Buck Rogers-style backpacks demonstrate space construction techniques erector-set fashion. Those allotted the special duty of a spacewalk (called in NASA lingo

"EVA" or extravehicular activity) must suit up for the hostility of the space vacuum as methodically and meticulously as a knight might have done in the Middle Ages. A spacesuit is, in fact, very similar to armor with its hard torso, cumbersome bulk, and protective headgear, and it must be donned with care.

There is only one way to get into a spacesuit. The astronaut must strip naked to his or her underpants. After attaching medical monitors, the spacewalker then puts on a liquid-cooled undergarment — essentially a set of long johns of very comfortable, loosely knit, elasticized fabric shot through with water hoses. Next come the pants and boots, which are of light-reflecting white Ortho fabric, underneath which are five layers of flexible metallic mylar for heat reflection (like mirrors they deflect light) and meteoroid protection. The suit crinkles when the astronaut bends or walks. Fabric in vital bending places like the fingers of the glove is pleated to allow for maximum flexibility and pressure equalization. The hard-shelled torso is next. Because the torso is not flexible the astronaut must place his hands over his head, as if diving, position hands and head in the direction of their respective holes, and then slowly bring both arms down simultaneously, a movement similar to the breast stroke in swimming. Once the torso is in place, it is connected to a metal ring on the pants that allows for limited turning at the waist. Last of all, helmet and gloves go on; they are attached through intricate locking devices in order not to break loose accidentally. The helmet is clear, allowing for nearly 150-degree forward visibility. The whole procedure takes at least an hour.

According to old NASA standards, the astronauts had to prebreathe oxygen for 3½ hours before egressing from the Orbiter, or risk getting the bends. Astronauts Peterson and Musgrave recall falling asleep in their spacesuits during the prebreathe because "there was nothing to do and it was the quietest place in the cabin." The long prebreathe was always regarded as a terrible waste of expensive manhours and so according to the new NASA guidelines the cabin pressure is dropped from its usual 14.7 psi to 10.2 psi the night before a spacewalk; the prebreathe then takes only about an hour to achieve the same amount of protection from the bends.

Spacewalking as it exists now goes against all the learned responses of human locomotion. Feet and legs are virtually useless.

In order to look behind oneself, one has to rotate one's whole body since the helmet is, according to Peterson, "like wearing a neck brace." It's hard to bend in this suit because it's cumbersome. Hand movements are greatly restricted in a pressurized glove. The fourth and fifth fingers are altogether useless, and because one cannot feel anything in the palm of one's hand, one must relearn all hand movements in order not to fight the glove. Reaching for an object with fingers outstretched is unnatural in a space glove and the hand tires quickly; instead, people must go back to a simian form of grasping, with fingers curled in a hook as if clinging to a tree branch. Tools must be specially developed with wide bases in order to be useful at all. Small delicate movements, such as using a screwdriver to turn a screw, are altogether impossible. Large vice grips are used for such activities instead. Fatigue results when an astronaut must hurry through an unplanned procedure. "When you try to hurry in a suit," said Peterson, "you really work a lot harder than if you go along at a nice steady pace."

On a spacewalk, sightseeing is limited by intense sunlight. The light is so intense that it is difficult to read instruments. And although one might expect stargazing during a spacewalk to be the greatest of all, the brightness of the sun obscures the stars. On STS-6, *Challenger* was allowed to do a ferris wheel orbit at the astronaut's request, so that part of the orbit would allow them to look directly out into space. "The thing that surprised us was that you couldn't see stars at all because there was so much light in the payload bay," said Peterson. "Your helmet picks up those lights, so your eyes don't dark adapt enough for you to see stars. We thought we'd be able to see a billion stars. If you got back in one corner of the bay with your back to the bay, you could see some stars, especially if the light was toward the back of your helmet. But you could do better looking at the stars on Earth or in the spacecraft."

During his spacewalk, Musgrave felt some sensation of speed: "I thought about how fast the Orbiter, the tools, and I were going backward and sideways down the road. Sundown was like turning all the lights off and sunrise was a sharp edge of color — very beautiful."

Even when an astronaut is not treated to the spectacular privilege of a spacewalk, there are plenty of strange and wonderful and bizarre events that change everyday spacelife into Disneyland. "I remember I liberated from an orange-juice container a glob of orange juice that was about as big as a coffee cup," said Joe Allen. "It was in front of me, and I moved it around as if it had come to life" (Allen & O'Toole 1983:114). He finally stuck a straw into it and drank it. Routine water dumps (the fuel cells always produce an excess of water that must be dumped) become a shower of prisms by day and a "west Texas snow storm" by night. The Orbiter might glow incandescent on its own with a space version of St. Elmo's fire. OMS rockets fire pinpoints of light while smaller vernier attitude rockets glow orange in the dark. The large RCS jets sound "like howitzers outside the window" or "like somebody has just jumped onto the roof of the Orbiter" while the smaller vernier jets make no detectable sound. During proximity operations, that is, when the shuttle is rendezvousing with another satellite, the attitude jets must fire and, according to Robert Crippen, commander of STS-7, "it sounds like a small war going on out there."

All the time they're working, the strange and splendid views allotted only to the privileged space travelers proceed past the numerous windows: the red-streaked sand of the Sahara, the innumerable greens and blues of the Caribbean, the subsurface mountains and valleys of the Pacific when the sun hits at a particular angle, the blue bands of the atmosphere at sunset, snow in the Andes and the Himalayas, the patchwork quilt of farms through the American Midwest, the wake and bow wave created around the Hawaiian Islands as strong currents rush by. On STS-7, while her crewmates were doing various hygienic and medical activities, Sally Ride turned off the lights on the flight deck and ate in the dark, "looking at the Earth in the moonlight," an almost mystical experience.

All is not fun and sightseeing. Some astronauts get sick with stomach cramps or feel uncomfortably hot or uncomfortably cold as the shuttle turns and changes its position toward the sun. The front windows are sometimes smeared from the chemical residue from solid rocket separation or fog up as a result of a poorly directed air vent. A switch on a cathode ray tube may have to be replaced and the fan filters have to be cleaned of debris frequently lest the air-cooled electronics overheat. The suction cup shoes designed for space walking "hold when you don't want them to hold and let go when you do want them to hold," said Bill Lenoir. "You just take off your boots and your socks and you can use your toes. It's like having four hands." Tightly packed food packets come floating out of the food warmers and have to be caught and stuffed back into the warmer. On some flights, a convection oven replaces the food warmers. Bob Overmyer needed to restrain his whole tall body in order to eat: "I got my feet up on the fourth step of the ladder on the middeck and my neck jammed against the bulkhead. I'm eating barbecue." The toilet is legendary for its malfunctions. The contours of one's derriere may not be such as to form a perfect seal with the toilet seat, in spite of foot restraints and a seat belt to hold the seated person in place, making it necessary for the astronaut to brace himself awkwardly. Apart from that, mechanical failures, such as the blockage of the suction air flow or breakdown of electronic components, occur with regularity.

Exercising on the treadmill takes time and causes discomfort. Sometimes astronauts generate enough sweat to cause a "small shower" in the middeck, and cleaning up is not easy. "It's trouble to get the treadmill set up, then get the harness on, exercise a half an hour, then restow the treadmill and clean up," said Peterson. "Cleaning up takes an hour. You have to be careful not get water free in the cabin, and you have to watch out when you lather up and sponge off. As a result you wind up with two or three wet towels and nothing to do with them." Trash from empty food containers, packaged equipment, used towels, and so on piles up to the degree that the astronauts have to strap it down to the middeck floor because there's no place else to put it. After a few days, compartments in which wet trash is stored take on a garbage odor.

Although most shuttle trips are very short, some astronauts do, consciously or unconsciously, think of home on Earth. Invariably, when the astronauts fly over Houston, they ask about the weather, the temperature, how the Houston Astros or the Houston Oilers are doing — essentially how life is going on below them. Occasionally, the more sentimental astronauts, especially during the early flights of the shuttle, felt moved to thank everybody — their families, their neighbors, Mission Control, all the workers in the space program — for the proud opportunity of flying the most advanced spacecraft in the world. People below them heartily returned their enthusiasm: families would be allowed into Mission Control to greet them. "Look down," said one astronaut's son, "and, I'll be waving!" Jack Lousma's neighbors erected one American flag each day on the median green for each day of his flight. Millions turned out for launches and landing, and journalists couldn't seem to get enough of the shuttle, the astronauts, and the romance of spaceflight.

Mission Control is always fastidious about passing along personal information — informing an astronaut of family news — if a child's temperature is down after an illness, if the kids' little league game was won or lost. In fact, Mission Control has its own special relationship — a combination of camaraderie, banter, trouble-shooting, professional advising, and ultimate guidance. Traditionally, a crew's first morning in space, the day after launch, starts with a wake-up call from Mission Control, often a song or a joke pertaining to some of the individuals aboard the flight. In the high hilarity of the early space shuttle flights, special songs might be composed in homage to the spaceborne from the Earthbound. For instance, STS-1 crewmen Young and Crippen heard an incongruously twangy country-western tune (the kind usually devoted to cheating hearts and lone little doggies) honor the space program.

> Well, many many hours went into this thing
> A job well done by the shuttle space team.
> We can't say she's sleek and lean,
> But I'll tell you right now she's a mean machine.
> The Columbia —
> Not the kind you smoke.
> This here's a bird.
> She gets high on herself.

Jokes were made about an astronaut's past embarrassing moments. "Let Crippen sleep!" screamed one good-morning message beamed up to STS-1, "Or he'll spill coffee all over the console again!" (a reference to when Bob Crippen knocked a cup of coffee onto the simulator console during a test run and shorted out the circuits). STS-5's Lenoir and Allen found inflated toy sharks packed into their spacesuits, which popped out and floated threateningly through the cabin. In some flights, the wake-up music included the "Cotton-Eye

Joe" for Joe Allen, the Texas A & M Fight Song for Aggies, the University of Texas Fight Song for U.T. graduates, the Marine Hymn for marines, and so on. Sometimes the crew brings along their own music to work by, and it's not unusual for sounds from the big band era, country-western ballads, and even the powerful classical favorites, like Richard Strauss' *Thus Spake Zarathustra* (the theme music to the popular film *2001*) to stream in through the speakers at Mission Control: the music spacepeople work by.

Russian cosmonauts once remarked that although they might long for the folks at home, they sometimes irritate them "with their continuous nitpicking." Houston's Mission Control always approaches unscheduled requests with a sense of decorum. A flight controller in Mission Control is a habitual worrier. He or she is trained to worry, trained to look ahead at every possible little thing that could go wrong. It is not unusual for such a person to request that a gauge be double checked or a number read back. Multiply that by the 90 worried engineers and it adds up to a lot of extra, unscheduled, nitpicky "it's probably all right" kind of tasks that the astronauts have to do over and above the already closely scheduled ones they are doing. As a result, Mission Control orders such requests by priority and asks the astronauts to carry them out with extreme decorum: "When you have time, could you. . ." "When you have a chance, we'd sure be grateful if you. . . ." "Don't let us disturb lunch, but afterward can you let us know about. . . ."

The decorum did break down during the flight of Spacelab 1 when eager scientists on the ground asked mission specialist Bob Parker and payload specialist Ulf Merbold to perform additional tasks while they were still trying to complete the scheduled experiments. "If you guys would recognize that there are two people up here trying to get all your stuff done, I think you might be quiet until we got one or the other of them done," said Parker testily. Spacelab 1 was the first flight in which scientists were allowed to talk directly to the astronauts in space, without having to go through an astronaut capsule communicator. Parker's sharp response reminded some of the last days of Skylab in which the astronauts were similarly irritated by unscheduled additional demands. Mission Control personnel took note and considered briefing scientists on future Spacelab missions on the value of precision in messages, appreciation of how long it takes to do a task in space, and radio manners.

In fact, crews sometimes savor the space experience and prefer to delay chatter with the ground. On STS-6, the crewmembers did not answer the hail from the ground until they had gotten their morning underway. Engineers in Mission Control could pick up signs of life — the flushing of the toilet, the activation of the food warmer, the turning on of the teleprinter. When communication between spacecraft and ground ceases at sleep time, the astronauts use some of that quiet out-of-contact time to linger over the view and enjoy the quiet. Often, they prefer to gather on the flightdeck with cups of coffee and talk as the Earth passes by below them — an informal gathering of friends in front of the most spectacular livingroom window imaginable. It is the only opportunity for them to reflect on the space experience, to absorb, and even revel in, the enlarged perspective they enjoy. "Sometimes, I wonder," said President Reagan in his conversation with the STS-5 astronauts, "if more of us could see it from that angle, we might realize that there must be a way to make it as united in reality here on Earth as it looks from outer space."

Sleep comes easily for some and with difficulty for others. Story Musgrave was so keyed up during his first day in space that he started the next day's work during sleeptime. According to crewmate Don Peterson: "Story kept us all awake for about 2½ hours going through the spacesuit checklist. He was like a 150-pound hamster."

Although there might be only two ways to sleep on Earth, sitting down or lying down, there are several ways to sleep in space. One can sleep strapped into a seat or can float freely in an open area; one can sleep on the flightdeck, the middeck, or in the airlock, with a sleeping bag or without one, with head strapped to a pillow or floating freely, with arms strapped to sides or dangling in front like a sleepwalker's. On some flights, the windows are blocked by sunshades so the sleepers on the flight deck will not be disturbed by the numerous sunrises as they orbit. On others, the windows are left unblocked and the sleepers wear eyeshades. "Who was that masked man?" Sally Ride said of her commander, Bob Crippen, who looked very much like the Lone Ranger as he slept in his seat, with mask on and arms folded "in the imperial position." In fact, on STS-7, no two astronauts slept alike. Crippen's pilot, Hauck, floated, unstrapped in his seat, while Fabian, Ride, and Thagard slept on the middeck, positioning themselves up on the lockers, down on the floor, or any way that was comfortable for the sleeper. Sally Ride slept cozily in a large sleeping bag with arms inside; her crewmates Fabian and Thagard slept in tighter sleeping bags with arms floating freely. On STS-6, Don Peterson found he had to anchor his head to the pillow with a small strap or "my head would bump into things and wake me up." Habitability specialists even developed a velcro pillow and headband so the astronaut could change his/her head position on the pillow without discomfort. Spacelab astronauts had to sleep in soundproof, coffin-like compartments because two shifts were required to work around the clock.

Sleep might be long and dreamless or might take some account of the space experience. Some astronauts dream of floating from place to place, as they do in space. Others dream of people and places they miss on Earth. Many do not dream at all, or rather, some recall dreaming but do not remember the content of their dreams. An astronaut might awaken during the night to a Halloween-like vision, as Bob Overmyer did: "Joe [Allen] slept kind of suspended near the teleprinter last night, in the eerie glow of the green CRT teleprinter. I was suspended in the middeck floating around. And every time I brushed against something, I woke up and saw this kind of body hanging over this greenish glow."

Homecoming involves a fiery reentry, with all forward windows glowing a deep salmon color, then lighter rose, then white. The astronauts are crushed down in their seats. The noise of the air rushing past their spacecraft can be heard, as crewmember's limbs grow heavier and heavier. During the night landing of STS-8, a pulsing wave of light was seen dancing around above the spacecraft, a phenomenon scientists still do not understand. But the night landing to observers on the ground looked even stranger: not lit up like an airliner, the shuttle appeared as a ghostly, lumbering beast, materializing just beyond the floodlights of Edwards Air Force Base.

In the early days of the shuttle program, homecoming meant tumultuous, cheering crowds; speeches by politicians and by the astronauts themselves; rounds of parties back in Houston; and months of public relations tours, including dinners with the President, receptions from the members of Congress, endless interviews — the whole pressing need of an entire nation to hear of the space experience.

Homecoming for the astronauts always seemed to be characterized by three emotions: a feeling of gratitude for having had the priviledge of flying in the shuttle, an appreciation for the Orbiter itself, and a sense of disorientation. According to Story Musgrave: "On your return, you are aware that there's a whole world of people who helped put you there. Also you have a real feeling for the Orbiter itself; it's not a hunk of metal. It has a real personality, and you've learned to care about it. Finally, you are disoriented. You have lunch in space, dinner on the ground and you ask yourself, did you really go at all?" Even during the pomp and circumstance of presidential dinners, congressional receptions, the frank admiration of 220 million Americans, the astronauts feel an inevitable sense of letdown. They are glad to return to work, look forward to their next flight assignments eagerly, long to be pulled from the emotional limbo, the calm aftermath of the greatest experience in their lives.

As the business of the shuttle became more ordinary, the crowds, mercifully, thinned, leaving only a few diehards who would never miss the beauty of a launch or the splendor of a landing. The interviews were fewer and briefer, the jubilation died down.

As early as STS-9, space had ceased to become headline news. More than two million people attended the launch of STS-1. All three networks and many independent stations sent anchorpersons to the Cape. Hours of television air time were devoted to all the various aspects of the shuttle, most especially the new and untried heat shield. By the time STS-9 was launched, only 50,000 attended the launch, the anchorpersons stayed in New York, and the evening news on all three networks led with a blizzard that had enveloped the Midwest. Major newspapers and weekly magazines began to send stringers rather than their prime correspondents to cover the space news.

The shuttle was no longer news, only a year and a half after the commencement of its operations. Ironically, the shuttle found itself, on the brink of its finest efforts, at parade's end.

The time was sooner than anybody, especially some astronauts, expected. When Joe Allen returned home from his first flight aboard STS-5, two signs greeted him. His children had written "Welcome Home, Dad" on one; his wife had written the other: "The lawn needs mowing."

2
What Kind of Person
Is a Spaceperson?

A few years after their historic moon landing, astronauts Michael Collins and Buzz Aldrin wrote biographies in which they confessed their own forms of mental illness. Collins confessed that he was a claustrophobic; Aldrin exposed his own bouts with depression. One writer quipped: "I wonder how Armstrong felt about riding to the moon in that tiny Apollo capsule flanked by someone who doesn't like cramped quarters on his left and by a guy whose moods are like a roller coaster on the other."

In fact, both astronauts had gone through the most rigorous psychological screening available at the time and performed well before, during, and after their flights.

Collins became claustrophobic only when he became warm in his spacesuit. He easily overcame the problem by having the coolant turned up to maximum whenever he was in his suit.

Aldrin's mental illness was of another order. While he was driven to achieve a goal, he controlled his depression well, although when he was alone, unmistakable symptoms did manifest themselves. Once the goal of landing on the moon was achieved, a professional vacuum developed in his life, rendering him directionless. In his candid book *Return to Earth* Aldrin described how his bouts with depression finally overcame him to the point where he required hospitalization (Aldrin & Warga 1973:295–314). The odd thing about his mental illness is that it was never detected by all the known psychological measurements of the time.

Early on in the space program, physicians and psychologists worried that the human psyche might not be able to bear the space environment — at that time, the great unknown. Human beings habitually resort to violence or hysteria when confronted by the unknown. There was no analogue to spaceflight in human experience. When Columbus was going across the sea to seek the riches of the East, he and his crewmembers would be sailing—something they were accustomed to. The same was true of land explorers like Marco Polo. But space was truly unknown.

A special Space Task Group, set up in 1958, looked at several categories of individuals who might be best at handling spaceflight: aircraft pilots, balloonists, submariners, deep sea divers, mountain climbers, explorers, flight surgeons, and scientists. They decided that the best candidates had to be graduates of test pilot school with at least 1,500 hours' flying time. These individuals were all used to danger, strange sensations, and the unknown. These candidates were then tested in a centrifuge, low-pressure chambers, planes flying in parabolic arcs, thermal exposure units, and so on. The duty of the finalists was simply, as Arnauld Nicogossian and James Parker, Jr., the authors of *Space Medicine and Physiology*, state: to survive, to perform, to serve as a backup for automatic systems, to observe, and to suggest improvements to future systems. "According to the Lovelace report the seven ultimately selected were chosen because of their exceptional resistance to mental, physical and psychological stresses and because of their particular scientific discipline or specialty" (Nicogossian and Parker 1982:234).

A flying background remained an absolute requirement for astronaut selection until 1978 and still remains a dominant factor in astronaut/payload specialist and ESA (European Space Agency) selections. Past good performance under life-threatening circumstances seemed to be a good measure of courage, or of what Hemingway called "grace under pressure."

Nonetheless, pilots can be great pilots and still have some latent psychological problems. The astronauts selected during the 1950s and 1960s were required to take the whole gamut of psychological tests involving personality, motivation, intelligence, and aptitude. The Mercury astronauts took a total of 23 tests, some of which were in multiple-choice mode, others of which were projective "inkblot"-

type tests, and others still that were not tests so much as written evaluations submitted by superiors and peers alike. These tests were only minimally useful to the early space program and were dropped in subsequent astronaut selections.

Astronauts are now given interviews with two NASA psychiatrists or psychologists in order to evaluate their mental stability. The psychiatrists might ask questions like: "What is the meaning of 'a rolling stone gathers no moss'" (a schizophrenic would say that it means that and only that — since he cannot abstract well); "If you could come back in another life, what would you come back as?" In the class of 1980, the favored response was "eagle." Exceptions included Mary Cleave, a Ph.D. sanitary engineer, who wanted to come back as a seagull because it was a useful, cleaner-upper kind of bird; another wanted to come back as an Air Force wife (a private joke).

In 1978 and 1980 selections, the psychiatrists played a good guy/ bad guy routine: one was a kind, compassionate, "let me be your friend" kind of interviewer while the other showed undue hostility and barked out the questions. This may not be the pattern for future selections. Interview styles will undoubtedly vary with the individual psychologists or psychiatrists NASA employs to do this work.

Some NASA psychologists would like to have some sort of neurometric assessments made as well, because these test a wide range of responses. "Psychometric evaluation [by written test and interview] assumes that the individual knows how he feels and is honest about his feelings," said Dr. Kirmach Natani, who has worked for many years with the Air Force helping to refine its own selection procedures.

Natani promotes a kind of evaluation called a neurometric measurement, in which involuntary nervous system responses are recorded during a number of tasks and tests. The equipment required for this is a desk computer, flow charts of procedures for testing tasks and responses, and a strange-looking test apparatus in which the candidate sits in a deliberately uncomfortable position as his or her physiological responses are tested. As the candidate's responses are tested, the physiological data are fed into the computer, which compiles and analyzes the information immediately, on a specific response-by-response basis. Since the test situations arise suddenly,

the candidate does not have time to deliberate. In this way, unconscious, nonverbal, innate responses to novel situations can be assessed. Such assessments could be used in the future, when space
travelers come from the public at large, not from stressed backgrounds such as those of test pilots. They may be used to weed out
those who are accident prone and those who become careless when
bored, qualities difficult to assess from written tests and psychiatric
interviews.

The Air Force has used a battery of neuropsychological tests to
select personnel for spaceflight. This series tests a wide range of
brain/behavior factors: how quickly a person assimilates new learning, how well a person conducts abstract reasoning, how well long-
term and short-term memory function, and so on. These tests look
not only for dysfunction but also at how well a person functions,
what psychological style a person possesses. Such tests are not yet
part of the NASA selection procedures but may become so in the
future, if NASA selection boards feel they need these to assess their
potential astronauts.

And then comes the rescue sphere. The rescue sphere is a pressurized beach ball affair dreamed up as part of a rescue scenario.
Say that one shuttle were crippled by some serious malfunction and
had marooned the astronauts in space. Say that a second shuttle
just happened to be on the pad awaiting launch. Hypothetically the
second shuttle could rescue the crew of the first, although evacuation of the crew would prove to be a dilemma since the marooned
astronauts would have to float across the vacuum of space. Since
there would usually be only two spacesuits onboard, some crewmembers would have to get across space in some other manner.
The answer was the rescue sphere. An astronaut would roll himself/
herself into the fetal position inside this pressurized beach ball and
could be carried, suitcase style, across space.

All the particulars of this scenario were somewhat farfetched,
although the rescue sphere did find its usefulness in the space
program — as a test for claustrophobia. Candidates were asked to
evaluate the rescue sphere and note their thoughts on its technical
feasibility. The person was zipped into the rescue sphere, in total
darkness, with cool air blowing in. Some became very relaxed and
even fell asleep; others asked, in some discomfort, when they could

be let out. One Air Force lieutenant colonel refused to get into the sphere at all and disqualified himself on the spot. "I guess you guys are serious about this," he said, "and I'm going to disqualify myself right now because I know that if I get into that thing, I'll go bats."

Some of the 1978 astronaut hopefuls were put through more than usual strenous medical tests, measuring strength, health, and so on as a part of an ongoing NASA project. Some bridled at the invasive tests; others tolerated them. Others yet endured them as cheerfully as possible. "They were, in some cases," said Jake Jackson who worked on anthropometric measurements, "a test of character. I remember one guy especially who had been in a car accident shortly before and had broken some bones. We didn't know about the accident and asked him to do some strenuous things, which he did try to do without a murmur of complaint, although he was in dire agony the whole time. I think that's a test of character."

Payload specialists go through a similar psychological evaluation. Charles Walker, the payload specialist flown by McDonnell-Douglas, had an interview with one psychiatrist and a session in the rescue sphere test.

Because past experience is considered the best predictor of future behavior, NASA might look favorably on future journalists, writers, artists, and so on who have some flying experience. Other "tests" might be required of these outsiders as part of the selection process for those applicants who do not fly and have never really been in a life-threatening situation. One training director thought that observing reactions of nonflyers during very uncomfortable aerobatic flying and KC-135 parabolic flight might give NASA an indication of how a common nonflying person might respond to spaceflight. Said journalist Dave Dooling, who had done a story about astronaut training in the KC-135: "You can get so sick that it makes you think about whether you really *really* want to fly in space that badly." Astronaut Don Peterson thought it might be worthwhile to put a potential spacefarer in a spacesuit and do some simple tasks in the water immersion facility pool in order to see how that person responds to a close situation in which he or she must rely on technology, must do things according to the limitations that the hostile environment places upon him or her.

The crucial factor in any astronaut selection is the final interview with the NASA selection committee consisting of NASA flight directorate personnel and veteran astronauts. The psychologists, physicians, and psychiatrists have, by now, weeded out the unsuitable candidates, leaving approximately 150 finalists (out of as many as 10,000 applicants in the 1978 selection and 4,700 in the 1984 one). Very few people are rejected for psychiatric problems. A few were found to be claustrophobic, a few others had an attitude problem that may have gotten in the way of astronaut work. But by and large, most are turned down as a result of medical problems: most for visual problems, some for hearing problems, a few others for physical defects such as a heart irregularity.

The talent pool that makes it to the final interview is really quite stunning. Most of the pilots have accumulated many hours of flying experience on numerous types of aircraft, have a distinguished combat record, and have been very exemplary test pilots — many of them are best of the year in their respective flight schools. Of the mission specialist scientists, almost all have Ph.Ds, or M.D.s, all have a trail of academic or professional awards, and many have remarkable hobbies. This cream of America's youth is not motivated by money, since the astronaut job offers a starting salary of only about $25,000; they cannot be motivated by glory, since the astronaut corps' only reward is a flight assignment, and even that may not be the particular assignment the astronaut wanted; and an egocentric genius might more suitably be established in a laboratory with a staff, since the astronauts are stacked four to an office and no account is taken of personal idiosyncrasies.

As head of flight operations George Abbey said: "These are all such remarkable people that they will be making *sacrifices* to get into the space program."

The interviewers are careful to scrutinize candidates for the ability to give 100 percent of themselves to the space program 100 percent of the time, to be able to work as a team member, and to handle emergencies calmly.

All the chosen astronauts have been deemed to have "the right stuff" over the years. What most people don't realize is that "the right stuff" of the space program changes according to the needs of the program. The early astronauts were picked because of their aggressive piloting skills, their cool reactions in the face of danger, their ability to deal with new situations and new aircraft. The fact that such individuals might be arrogant, who regarded the NASA assignment as only one more hot glory after which they would push off and do their own thing was not deemed important. Neither did NASA have within its means the usual methods of control for such individuals at the time: a large bureaucracy, a seniority system, the ultimate threat of denying a person a flight assignment, and so on. It would be unthinkable for one of the current shuttle astronauts to face down an American vice president, as John Glenn did with Lyndon Johnson, or for an astronaut to insist on cetain spacecraft modifications as all the original Mercury astronauts did.

As the space program grew through the Gemini and Apollo days, NASA came to realize that while it was okay to have competitive astronauts, it was a distinct problem to have contentious ones. While it was reasonable for an astronaut to be proud of the profession, it was unprofessional to be arrogant. It dawned on flight operations people that the solitary trailblazers who took us into space, into orbit, and to the moon might be unfit by temperament and too restless by nature for the next stage of space development: the scientific investigation of the new frontier and the long-term occu-pation of it.

First some scientists were allowed into the former all-pilot ranks of the astronaut corps. In 1965 and 1967, seventeen scientists were selected to become astronauts. *But they all had to learn to fly.* The pilot mentality had so dominated NASA that it was inconceivable that a nonpilot might be able to do well in space. During the moon-landing era, the pilot mentality still dominated the astronaut corps to such a degree that only one scientist, geologist Harrison Schmitt, was allowed to land on the moon, and that only after a skirmish within the NASA hierarchy. All the others were pilots, trained in geology it is true, but not professional geologists.

Although half of the shuttle astronauts are professional scientists now, scientists with pilot licenses or background are still heavily favored. "The best analogue to spaceflight," said Alan Bean, former astronaut and one member of the selection board for the 1978 and 1980 candidates, "is aviation." According to pilot Roger Zweig, an astronaut instructor, a plane teaches its own kinds of lessons: "Even nonpilots must learn about flying in three dimensions — the sensations, the discomforts. It teaches crew coordination by teaching teamwork. It accustoms people to using displays and having to worry about consumables."

Ironically, the great "stick-and-rudder" fliers of the early space program would be a liability to the shuttle. According to one test director: "One older stick-and-rudder-type astronaut killed himself four times in one day in the shuttle simulator. His embarrassment taught him that doesn't work anymore."

In the 1978 and 1980 astronaut selections, nonfliers were invariably asked if they had ever been in a position of danger. One nonpilot, Jeff Hoffman, is a parachutist. He told the interview board that once his parachute only half deployed. He calmly calculated that his rate of descent was too fast to survive. He deployed his backup chute only after cutting loose his main chute, since the one could wrap around the other. The step-by-step, cool, almost analytical way in which Hoffman assessed his situation and then weighed alternatives must have pleased the interviewers.

Astronaut Rhea Seddon focused attention on one other trait, perhaps the most important to current astronaut selection boards: "We are no longer the solitary warriors of the early space program; now, we are more team players." In fact, NASA's experience proved that solitary warriors were somewhat incompatible with the new needs of the space shuttle program — the close quarters, the close cooperation necessary to perform some of the experiments, the personal discomfort, spartan living conditions, and the inherent complicated interreliability of the shuttle systems. In order to launch a satellite, one crewmember monitors the controls of the launch system, another documents the operations with a camera, another goes over all the checklist items, while yet another watches the computer displays so that the attitude of the spacecraft is kept stable and correct. In essence, four people have to coordinate their efforts smoothly in order for one routine function to go off properly.

Even after initial selection, NASA likes to give itself the option of weeding out unfit individuals who may have made it through the selection process but who did not fit in on the job. The 1978, 1980, and 1984 selectees were not formally inducted into the astronaut corps until their one-year initiation period was over, during which they were called "astronaut candidates." That initiation period included rigorous survival training, difficult classroom work, strenuous aerobatic aviation in a T-38, flights in KC-135s, and numerous other experiences that demonstrated to NASA their fitness for spaceflight. Their performance was watched and assessed the whole time.

It is possible that guidelines could change further with the ongoing needs of the space program. Some psychologists feel that a large number of people can put up with the rigors of spaceflight for a week or so. But when a space station becomes operational, new psychological pressures will be placed on the spacefarers, and yet another kind of astronaut, one adapted to long-duration spaceflight, might arise. Astronauts adapted to short-term spaceflight might find long-duration flight too boring, less challenging — unless their innate professionalism and their tireless curiosity keep boredom from ever happening. (I hasten to add that no astronaut I have ever talked to can even *conceive* of being bored with spaceflight, ever.) Vladimir Lyakhov and Aleksandr Aleksandrov, both long-term space veterans, valued two human qualities above all others in spaceflight: "Professionalism and kindness. In life it rarely happens that a good professional is a bad comrade. In cosmonaut selection both a person's character and his special training are taken into account. In none of the professions I know is there such a careful and comprehensive analysis of each person."

Long-duration flights like the American Skylab missions and the Russian Salyut missions have been not only experiments in discovering the scientific potential of space but also the testing ground for finding the psychological limitations of spacefarers. The head of Russian space biomedicine, Dr. Oleg Gazenko, said bluntly that "the limitations to man's living in space is not physical but psychological. All the young cosmonauts want spaceflights to become longer, but the older ones who have been in space for long-term flights don't like it. Ryumin [the only person on Earth who has spent a total of almost a year in space on his two trips on Salyut] hates spaceflight."

Space is a sensory deprived, spatially restricted, isolated environment in which to live. "One is always sharply aware that one is not on Earth during a flight," said veteran cosmonaut Sevastyanov, "that one is separate. And this is not at all the sensation one feels when one is on Earth, shut up in the isolation chamber even for a long time. No, I was always acutely aware of being remote from Earth while I was in space."

According to some accounts, the Skylab IV crew went on strike while in space owing to the large amount of extra work asked for by the ground, added to a mountainous amount of work already scheduled. Sevastyanov also complained of the "constant nitpicking" by ground controllers. Dr. B. J. Bluth, a sociologist who has studied space interactions, said that one cosmonaut crew shut off their air-to ground radio link for a day, and another time, two cosmonauts actually got into a fist fight in space. Although these stories have never been officially acknowledged, it is known that a privacy curtain was sent up on a Progress supply ship along with food, mail, and spare parts.

Apart from hostility, several other symptoms accompany long-duration isolation tours in space and on the ground. Insomnia, carelessness, and depression can all manifest themselves. To a degree these are life threatening.

Valentin Lebedev noted in his diary two months after he had been in space: "Against the background of increasing tiredness, the blunders in communication, in the work with Earth, lie in wait; and there are tense moments between the crew, but no outburst can be permitted. Otherwise, if a crack appears, it will get wider." After three and a half months in space, Lebedev wrote: "I started to sleep badly. I lie and dream about different things, and I remember home. I fall asleep somewhere around two in the morning." The night before he landed, nearly seven months after his launch, he wrote: "A mood I do not understand: worry. How are things down there? Our life has been adapted to a small little island in space, and then suddenly, the Big World! I am not myself.... Go to sleep!"

Several NASA-sponsored studies in the psychology of spacecrews are currently going on, to head off problems that might arise in some future American space station. There are several areas of investigation for space psychologists: isolation/confinement studies, sensory arousal, group interaction, man/machine interaction, and psychophysiological studies, including stress and biorhythms. Although no single study will render up a formula for psychological health in space, the various aspects of man's responses to space (on a social, motivational, physical, and sensory level) may help form guidelines for good space station design, sound rules, and reasonable behaviorial guidelines.

Two studies currently being done at Ames Research Center are surveying all the possible analogies to space, one with an eye toward group interaction and the other toward human behavior as it might impact space station design. Another study, at NASA Headquarters, looks at submarine duty as the closest analogy to space station life. Other researchers are looking at personality theory, questions of achievement and motivation, in future selection procedures.

It is difficult to study future space station situations in Earth-based university laboratory experiments; however, such studies may have some limited usefulness. One such study, carried out at The Johns Hopkins Hospital by psychologists Joseph Brady and Henry Emurian, placed college-educated subjects, three at a time, in a small space for 10 days to test group interaction under confined and isolated conditions. Contact with the outside was made through written messages on a computer. A large number of tasks, some involving close group cooperation, were given the subjects, who were rewarded or not rewarded according to their performance. In spite of stringent psychological screening procedures to keep out unstable personalities, one test subject had to be pulled out for hearing voices.

The students had to rate their feelings and moods during their confinement. Not surprisingly, on days in which adverse circumstances prevailed, the subjects were more hostile with one another and with the outside researchers. One damaged equipment while another went on strike. Sometimes the group of three fragmented into an "in group" pair and one social isolate. When a woman was put in with two men, the woman tended to pair off with one of the males, although performance of the group was not significantly altered from results of other groups. Brady and Emurian found that rewards were better motivations for good performance, that cooperation fostered better individual performance than competition, and withdrawal of one member and replacement with a "novitiate" tended to have a temporary negative effect on group performance. The study did not try to factor in differences in personality, since it was designed along strictly Skinnerian behaviorist lines, that is, that environment outweighs all else. Studies of this type are limited in their usefulness to spaceflight — because in space, complex psychophysiological processes are at work, different types of people are involved, and different group dynamics reign in crews that have been together and trained together for a long time. They do, however, study problems of crowding, motivation, conflict, and boredom, all important components of space work.

Antarctic expeditions, submarine duty, and oil rig assignments do offer some analogues to spaceflight. None of these are perfect, since they do not have to contend with the adverse physical effects of weightlessness, the tight scheduling of experiments, and so on. According to Dr. Marianne Rudisill, a habitability scientist working for the Lockheed space station group, "technical isolates" fare best in the Antarctic but may do poorly in space because of the very tight living conditions. On the other hand, communications between the ground and the spacecrew are excellent and almost continuous, and so spaceflight has the advantage over both submarine duty and Antarctic expeditions. However, psychologists are still interested in group dynamics in these isolated occupations because they serve as the only Earth-bound analogues to spaceflight and may yield up lessons that might be useful on future space stations. (One is that alcohol is extremely destructive, the catalyst to many violent catastrophic episodes in the Antarctic. The present guidelines of no booze in space will be continued.)

Psychologists can improve living conditions on space stations by working closely with spacecraft designers. According to Marianne Rudisill, past space station crews have pointed out that design details can be aggravating. She points out, for example, that the Skylab astronauts complained that the color of their outfits got to them after a while and that a change would be good. Furthermore, the clothing has to have the right kind and number of pockets, must be able to stretch to conform with changing body dimensions in space, and must have cuffs that cling rather than float freely, or worse, ride up. Food is very important, and they requested that "more filet mignons and vanilla ice cream" be added in later flights. A more convenient food service will have to be developed for space stations — one that is quick and easy, allows for a great deal of variation, and does not generate a great deal of refuse. Windows are also important as a psychological release, since looking out the window was the pet pastime of the Skylab astronauts.

She and other habitability experts have studied a wide range of problems, including food, clothing, tools, and man/machine interaction, in order to help spacecraft designers arrive at a sound, habitable space station design that takes into account the people who must function in that environment. Waste disposal, hygiene problems involved with shaving and bathing, and impractical design or placement of equipment or tools dominate astronaut complaints.

NASA may want to study the possibility of sending married couples on a long space station assignment. "It tends to foster jealousy in other people and undermines the cohesiveness of the group that is based upon their mutual survival," said Rudisill. "It is precisely because married couples see themselves as a pair and may be accused of taking sides in controversies because of their loyalty to their mate." On the other hand, some psychologists feel that a childless married couple might be ideal candidates for space station duty since separation from loved ones — a big space "downer" — would be eliminated.

Some lessons gleaned from Skylab and Salyut may be applied to future space station crews since both of these were on-site laboratories for human behavior in space.

1. Like the Antarctic expeditions, which found that "all people are equal in their common struggle against the elements," so also Russian commanders and crewmen alike shared in the unpleasant housekeeping tasks, avoiding unnecessary hostility and bad feelings associated when one person pulls "latrine duty" for the duration. On shuttle flights, whoever is not in the midst of some task might fix the meal for everybody. However, important judgment decisions, as on a naval vessel, are made by the commander. In the future, in all but emergency situations, this may change, especially when military or scientific payload specialists have the last word on the well-being of their experiments or when a space station gets a larger population — say, more than four or six. "With space growth, thinking in the kinds of leadership must grow too," said Treive Tanner, an Ames contract monitor who oversees many of the ongoing psychological projects. Styles of leadership are already an area of study in the Air Force and may be useful in a space station as well.

2. Close communication with Earth is a double-edged sword. Space travelers talk to their families during a flight, and the Russians even set up a two-way television system so that the cosmonauts could see their families as they spoke to them. In fact, every Sunday, the relatives and children of space station crewmembers come into the control center. "The conversations," said *Pravda* correspondent Vladimir Gubarev, "are always moving; they recharge the cosmonauts for the next working week. Like the medical observations, the schedules for these meetings are observed meticulously during the flight." These meetings also enhanced, however, their sense of remoteness from Earth: "Tears flowed when we heard their voices," recalled Sevastyanov. But he felt that the overall psychological effect was "cathartic"; in essence, it was good to have "a good cry," since it released accumulated mental stress.

3. Work is highlighted. Quite often cosmonauts and astronauts act as simple laboratory technicians carrying out the scientific experiments of others. The American astronauts accept this as the necessary "generalist" training all astronauts get, but the Russians found that a tremendous psychological boost was experienced by their cosmonauts when the scientists told them the results of their experiments after analysis on Earth and asked them for their own creative assessments in modifying and improving future experiments. The sense of being an important functionary and of doing useful work is sometimes lost after several months of running experiments; but being an active participant seemed to enhance their perspective on their work. Certain tasks, like spacewalking, will always be wonderful work. According to Valentin Lebedev, his spacewalk "was worth all the years of hard study, the agitation, the experiences, the sweat. My first feeling on opening the hatch was a huge Earth and a real sense of the unreality of everything that was happening. Space is very beautiful: The dark velvet of the sky, the blue halo of the Earth, and the lakes, rivers, fields and cloud formations speeding by. All around there is perfect silence, no sense of the speed of the flight. No wind whistles in your ears, nothing weighs on you. The panorama is calm, majestic. The station is frozen in space like a block. The station hatch shone like an open door in a house in the countryside. All around was the dense blackness of the space night."

4. Even space voyagers seem to need a day off. On short-term flights, it is possible to work long hours without too much time for relaxation, but not on a long-term space mission. Skylab IV astronauts worked 16 hours a day with no day off. Additional work piled on them encouraged a sense of malaise. Russian cosmonauts get a day off every week, a schedule similar to that worked on Earth. As Henry Cooper said in his

book *A House in Space:* one of the important lessons was that it was important for people to have the time to absorb spaceflight as an experience (Cooper 1975).

5. Do not disrupt those circadian rhythms. At first, people thought it might be best for spaceworkers to work in shifts so that someone would be awake all the time. Experience soon proved that shift work interferes with the circadian rhythms, that sleep is disturbed by the activity of people who are awake, and that shift working is lonely, since some companions are asleep. Skylab crews found that they worked best when all dined together, worked together, and went to bed at the same time. The same is true of the Russians, who try to maintain as "Earthlike" a schedule as possible.

However, on a large station, it might be feasible to run two 12-hour shifts, as was done on the recent Spacelab 1, with no ill effects. In that case, thought has to be given to separating the sleepers from the activity aboard the station.

6. Gardening and looking out the window are pet pastimes. At first engineers thought of not putting a window in Skylab at all. However, wisdom prevailed, fortunately, and the Skylab astronauts found windows to be a necessity, not a luxury, of space living. Colors are vivid, vision is acute, and the perspective is magnificent — one of the only nice things in a sensory deprived place like a space station. Russian cosmonauts also found that they enjoyed the plant experiments aboard the station: growing peas, green onions, and so forth. They even took their free time tending the little plants lovingly. Lebedev recalled: "Peas grow in a very interesting way: they come out of the soil already robust, with a thick stalk and leaves, and they are densely packed; wheat just goes straight up, like a green ray of light. It is pleasant to touch the shoots with the palm of the hand — they tickle." Gardens in space stations have been widely discussed in the habitability community. Although the earliest American space stations will not have one, it is likely that subsequent stations will, to augment boring prepackaged space meals with fresh produce as a psychological respite.

In future, more psychological studies on habitability will guide space station designers. Machines will take over some tasks performed now by astronauts. Displays will be designed with color coding, so that the astronaut need not look at every line in order to find the information wanted. Equipment will be more closely designed to interface with the people. A meeting area large enough to hold all the crewmembers might be important.

Intense studies in group dynamics may render a profile for future space station inhabitants. It is possible that candidates with isolation tours of duty, either submarine or Antarctic, might fare well, since their experience gives personnel boards a data point on how they will behave, just as aviation gave personnel boards of the past a predictor for spaceflight. Spacelab flights are sure to give psychologists some data on how shift scheduling works with a large group of highly motivated scientists.

The Russians, who give psychological support high priority in their program, may share their psychological insights with us. Russians have already stated that three months is the best space station tour length; crew efficiency is at its peak. Durations of 30 days and longer than five months are less than optimal because of physical adaptation problems in the former and exhaustion problems in the latter. The Russians punctuate these long-duration space voyages with one-week visits from visiting cosmonauts (which give an emotional boost although they also raise stress and reduce work efficiency). There are occasions for clowning, a change of pace. Unmanned capsules periodically fly up with fresh supplies, including fresh foods, little packages, all-important letters, newspapers, and once even a guitar.

"Space is a double loneliness," stated Lyakhov. "We are living here and in that case we experience all human feelings. Another thing is that you cannot 'go on a spree.' The main thing is to fulfill the program and this means work, work, work.... For us a letter is an extraordinary event. We read them over and over many times. Letters from dear ones, relations and friends."

Svetlana Savitskaya, the second woman in space, suggested that a psychologist be included on some future long-duration flight, to observe first hand the individual stresses and group dynamics. Dr. Patricia Santy of the University of Wisconsin Medical School brought up that exact suggestion in her article in the *American Journal of Psychiatry*: "The psychological monitoring of humans in space requires that behavioral scientists should be part of the crew in order to conduct continuing research into man's psychological adaptation to space exploration. Astronauts have routinely reported on subjective experiences and have kept detailed logs during the missions.... trained observers and researchers might provide even more reliable information and insight. The psychiatrist does not have to be considered a threat to the rest of the crew or as someone invading the privacy of crew members . . . One possibility is to involve psychologically trained mission specialists in making more detailed and ongoing recordings of astronauts' emotional responses to various positive and negative stresses in space . . . . brief psychological tests could be administered at specific times" (Santy 1983:525).

There is evidence that space changes people's perspective, sometimes profoundly, sometimes subtly. Among the people who went to the moon, one is now a religious zealot, one is involved with ESP possibilities, one is an artist, while others are engineering consultants, and one is still an astronaut. Their responses to their experiences on the moon impacted them differently, and it is possible to think of psychology beyond the Earth as the next vista in understanding psychological evolution — not just as man responds to space but as mankind is changed, as man evolves mentally beyond those responses he brought with him from Earth. As Santy stated, the area of "psychogeography" may develop as a separate realm of psychiatry: "The problem involves nothing less than the concept of separation/individuation of our species from this planet, Mother Earth. Space exploration and the colonization of space by humankind offers a unique opportunity to study the motivation of our species as a whole."

An example of this can be seen in Gerald Carr's experiences. Carr commanded Skylab IV, an 84-day mission, the longest to date of any American mission. Spaceflight gave him a new perspective: "From space, there are no national boundaries like on a map," he said. "Being out there, looking back at Earth, washes away notions of nationalism and provincialism."

Unexpectedly, his unconscious mind made the adjustment to space, too. Carr kept a diary in which he recorded his feelings, the things that occurred, and special events during the voyage. On day 50 he recorded his first zero-g dream: "Before day 50 I had 1-g dreams in which I moved and behaved pretty much as I did on Earth. On day 50, I had my first zero-g dream in which every action took place as it would in space. It was as if my unconscious mind had let go of Earth."

Once back on Earth, Carr recalled spending a restless night between 1-g and zero-g dreams.

On the other hand, Vladimir Lyakhov, another long-duration space traveler, dreamed of home, as of Paradise: "I dreamed I was fishing. There was a fine dawn and I hauled in a fish weighing 9 kilograms 200 grams [about 20 pounds]. About a year before the launch Sasha [his son] and I went fishing. He helped me to land this fish. And without a net."

3
The Astronaut Job

When new astronauts show up for work at the Johnson Space Center for the first time, they have already won many competitions. The mission specialists, most armed with Ph.D.s or M.D.s, have performed splendidly academically; pilots have come out on top of the most competitive flight schools in the country. Beyond that, the newcomers have undergone what is considered to be the most selective and possibly most thorough job application program in the nation and have, in short, qualified for the most elite job in the history of mankind.

But as soon as they drive past the giant Apollo-Saturn rocket on display, past museums proudly showing relics of the past space program, past the many offices where the future glories are already in progress, they are the greenest of green initiates.

Over the next two or three years, astronaut training will convert the way they act, react, and perceive.

The space environment calls for a generalist, someone who is intensively interdisciplinary. Survival depends on knowing all the systems of one of the world's most complex machines. The work is, by definition, exploring the unknown. Astronomers who have Ph.D.s in gamma ray astronomy must learn computer operations; those with Ph.D.s in theoretical physics must learn mechanical engineering; Ph.D. biochemists must become knowledgeable in celestial mechanics; physicians must become masters of dials, displays, and switches; geologists must learn human physiology. Beyond that, the astronauts must serve the American people; they must become skilled at public speaking and at handling a contentious press corps.

All-important reflexes are first hammered and then fine tuned into those appropriate for the shuttle. In an emergency, the astronaut — no longer a civilian scientist or military test pilot — will override any earlier reflexes and do what he or she must do, without thinking.

"It's a growing, expanding kind of job," said Guion Bluford, the first American black astronaut.

During the first year of training at NASA, the new astronaut isn't even called an astronaut. Rather he or she is designated an "astronaut candidate," a term somewhat reminiscent of "Ph.D. candidate," indicating that there is still an arduous path ahead. Within the astronaut corps itself, there are further delineations: one is considered an astronaut by them only when one has flown in space. At the very top rung of the ladder is the head of the astronaut office, currently John Young, the most experienced astronaut in the world.

In the first few weeks, the astronaut candidates must go through a kind of military basic training in order to be able to survive an ejection from a T-38 jet trainer. Parachuting, parasailing, and water survival are all part of the curriculum, and the lessons are strictly learn-by-doing. They are taught to get into a raft and to be picked up by a helicopter. They are taught to get out from under their parachute if it happens to settle on them during a water landing. They are taught the proper procedure for landing on hard ground during a parachute jump.

Then come the arduous weeks of classroom work covering the basic sciences involved with the shuttle: electronics, computers, aerodynamics, orbital mechanics, spaceflight physiology, Earth observations, astronomy, planetary science, space physics, atmospheric science, materials science, and star identification. After technical theory comes familiarization with spacecraft design and operations, including overviews of ground support activities and mission-planning procedures. They then study all aspects of the shuttle, just as medical students learn about the human body, system by system. The shuttles' body has its own anatomical and physiological information. One by one the shuttle systems are studied and the terms memorized: structure, propulsion, hydraulic and auxiliary power, thermal protection and control, environmental control and life support, avionics and data processing, guidance and control hardware, communications, instrumentation, tracking techniques.

Little by little, the language of space seeps into the astronauts' mind and begins to pervade their conversation: FOD (Flight Operations Directorate), RMS (Remote Manipulator System), MMU (Manned Maneuvering Unit), FDF (Flight Data File) are scattered generously in their discourse with other astronauts. They are learning the language of space.

While to the outsider, this type of training might seem unnecessary (after all, one can drive a car without learning the jargon or understanding the processes of automotive mechanics), the astronaut must speak to the shuttle in its own language, and the Orbiter responds with numbers and displays. It is the first step to appreciating the machine, working with it as a partner, rather than driving it as a master. In order to drive a car, one merely has to learn a few maneuvers — how to stop and start it, how to turn the wheel, how to accelerate and decelerate. But the shuttle is a little like the mystical dragons of medieval tales. To communicate with it, one has to learn its language; to get it to respond, one must spend years training to be prepared for its idiosyncrasies. This requires an understanding not only of what correct buttons to push but why, and what implications these actions might have on other systems of the machine.

All along, the new astronauts listen to crew debriefings from recently flown flights — critiques of how the flight went, problems that came up, and so on. The whole time, they are assigned a certain number of hours' flying in the T-38 jet trainer and become used to aerobatic flying and aviation technology if they are not already. Guest speakers, experts in their fields, lecture the astronauts on related fields of interest like the Russian space program and planetary geology.

The first year is a little like school, a little like military training, a little like on-the-job training, but it not exactly like any of those. There are no tests, no grades, no lectures from the teacher; similarly there are no praises, congratulations, or browny points. Highly competitive people who have excelled in the hierarchical worlds of academia and the military find the rules changed here. "The astronaut corps is a different king of pyramid structure," said former astronaut Alan Bean.

"It's not like flight school and it's not like school. Consistent good performance counts, of course, but so do other 'intangibles.' These people are all very competitive, but they can't be blatantly so. They have to be able to accept intellectually and emotionally the fact that other people have as good or better ideas than they do. Political astuteness doesn't mean one kisses up to the power. Instead, a politically astute person grasps the rules of the group and goes with them," concluded Bean.

Ten months to a year will have passed from those first footsteps onto Johnson Space Center soil to graduation day, when the newcomers are called "astronauts" and are pronounced "ready for flight assignment." They have lived, breathed, and dreamed spaceflight and have done so willingly, since they are all, as Frank Hughes, head of astronaut training, has called them: "hyperdrive people." By this time, the specialists have left their specialty behind, although some struggle to maintain certain professional skills, especially physicians who have to keep up their surgical proficiency. They are slowly becoming accepted as members of the "astronaut club," grasping the intangible and unspoken rules that will determine their future flight assignments and perhaps their entire professional life.

And yet, theoretical knowledge of the shuttle is not enough. They still lack, according to astronaut Story Musgrave, "the right reflexes because this is not driver's education in which you just do a few simulations and fly." Real knowledge is still far off.

Two to five years of "technical assignments" follow the initial year of astronaut candidacy. The class of 1978 had to wait five years before the first members of that class were allowed to fly. (There was a long list of older astronauts who had waited fourteen years or more for the chance to fly.) The astronaut class of 1980 had to wait only four years before its members began to fly. "The astronaut class of 1984 will have to wait only two, at the most three, years before they get their chance," said Hughes.

Each technical assignment is a kind of initiation rite, a new environment in which the astronaut must learn to function effectively. The Johnson Space Center is not a school but an operations facility. The many facilities do not exist solely to train astronauts but rather have their own ongoing work, into which the astronaut must fit. However, most technical assignments involve "hands-on" experience in which those important "reflexes" are acquired.

All the 1978 and 1980 astronaut classes were assigned to SAIL, the Shuttle Avionics Integration Laboratory, — a little-known but crucial facility at the Johnson Space Center. If outsiders were allowed to tour SAIL (which they are not), they would be treated to a view of the guts, sinew, muscle, tendon, and nervous system of the shuttle: the frenzy of wires and circuits that lie behind the clean white panels and under the metal floors.

SAIL is housed in a large building at the Johnson Space Center that virtually hums with the cosmic OM of computers day and night. All combined, the computers sound somewhat shrill, a cross between a jet engine and a vacuum cleaner. SAIL's job is to test much of the electrical hardware and some of the computer software on the shuttle. The SAIL personnel review test procedures, chase "anomalies" (improper electrical messages) through the tangle of wiring, and simulate electrical failures of every conceivable kind. Since the shuttle is one massive complex of wires, SAIL's work is long, tedious, exacting, and very precise.

Fifteen astronauts are assigned to work in shifts around the clock with SAIL personnel. Their job is to sit in the exact duplicate cockpit, use the controls, and read the display screens as the various electrical problems are simulated. Videotapes are made in the cockpit during tests, to monitor the astronauts' actions and to record the displays on the screens.

In SAIL, astronauts learn about the complex shuttle control system through "hands-on" experience; when emergencies are simulated, "the adrenalin can really start running," said astronaut Bonnie Dunbar. Their reflexes are trained to be swift and sure. And they get an appreciation for the shuttle's electrical peculiarities. They can get a "feel" for the spacecraft's quirks and learn to coordinate their efforts with those of other astronauts, getting a taste for situations that may come up in real life.

SAIL is, as head of astronaut training Frank Hughes said, "a showcase of astronaut talents. It is a good but inefficient trainer because the main purpose of SAIL is to test avionics, so you might only do two launches in a eight-hour shift." In future, SAIL will still require some astronauts to work the machinery, and all payload hardware and software will still be tested and verified there. However, it is unlikely that all members of future astronaut classes will wind up being assigned there full time.

However, all astronauts, past, present, and future, are required to spend time flying in a T-38. The T-38 jet trainer has been around since the early 1960s but is still favored by flyers for its responsive maneuvering, its acrobatic capabilities, and its efficiency. Pilots use it to keep piloting skills sharp, but it is an ideal teaching machine for nonpilot astronauts, too. "It teaches nonpilot astronauts certain intangibles," said Roger Zwieg, a NASA test pilot responsible for astronaut flightline training. "Nonpilot astronauts get a feeling for working as part of a crew. They get accustomed to cockpit controls and displays. They get used to real 'seat of the pants' flight, which no simulator can ever duplicate—the up, down, and upside down of real three dimensions. And they learn to manage their resources, appreciate the problems in dealing with consumable fuels that can run out." As astronauts fly in aerobatic maneuvers, they are taught the initial unpleasantness of pulling gs, the distortion of vision and of the senses, the changes in perception, sense of position, and attitude. Some astronauts are convinced that aerobatic flying before flight fortifies the astronaut for the stomach-wrenching effects of launch, entry, and zero gravity.

On an average day, six or eight T-38s might be lined up like so many rental cars on the flight line at Ellington Air Force Base, which adjoins the Johnson Space Center. Of all astronaut training, the astronauts enjoy T-38 the most.

The clumsy KC-135, affectionately known as the "vomit comet," is another aircraft in which astronauts spend time.

Essentially a huge flying laboratory, the KC-135 flies in parabolic arcs simulating two gs at the arc's bottom and zero g at the arc's upper half. The zero-gravity phase lasts only about 30 seconds, but in that time, everything that's not hammered down in the cabin floats up. Quite often equipment for possible use in space is briefly tested, and quite often human physiology is tested as well. Astronauts being taught to do extra-vehicular activities (EVAs) must learn to wiggle into their spacesuits in the KC-135. Astronaut Story Musgrave recalled learning a great deal about spacesuit movement in KC-135 tests: "Getting dressed in the zero-gravity portion of the parabola was a problem because I was part in and part out when we went into the two-g phase of it. But I learned that if you relax, the spacesuit will take you somewhere; it has a mind of its own. Also, there's no really good test for claustrophobia except a spacesuit—you get grateful if you get a little air to breathe."

Simple equipment, including a sink in which astronauts can wash their hands, is first tested on the KC-135. Some Earth-bound skills are sometimes tried out to see if they would work in space. For instance, when astronaut physicians were trying to find out if CPR (cardiopulmonary resuscitation) would work in space, they first tried to simulate the procedure in the KC-135. It was a disaster. The patient floated, the physicians floated, and it was almost impossible to get enough pressure on the chest to pump the heart. Instead, a device was developed that was strapped onto the patient and mechanically thumped the chest.

Again, some astronauts believe that KC-135 flights fortify the stomach for space flight. On the other hand, physicians point out that some astronauts who never got sick on a T-38 or a KC-135 did suffer from SAS (space adaptation syndrome) in space while some who did get sick in the KC-135 did fine in space. The effectiveness of frequent KC-135 flights to offset severe cases of SAS is, at this point, debatable.

Pilot astronaut training differs significantly from mission specialist training since the pilots are involved with the actual flying while the mission specialists are responsible for carrying out vehicle and payload-related tasks.

Apart from flying the T-38 many hours a month, pilots are rotated through the numerous trainers and simulators, where they get hands-on experience with each system of the shuttle: the guidance and navigation simulator, the remote manipulator system, the systems engineering simulator, and—much later—the shuttle mission simulator, the motion base simulator, and the shuttle training aircraft. Pilots who have been great stick-and-rudder persons in the past must learn to be computer experts. They have to work through checklists of procedures carefully and must learn to deal with the magnificent autopilot, which keeps a position in space automatically, firing an assortment of the shuttle's 38 jets whenever necessary to maintain an attitude. In effect, they must learn completely different instincts. Specialized pilot tasks, like proximity operations (which allow the shuttle to rendezvous with an object in space), require long hours of training and a thorough understanding of the attitude, propulsion, and guidance systems.

The shuttle training aircraft (STA) is an airplane that has been aerodynamically altered to fly somewhat like the shuttle by adding two large structures to the bottom. In addition to that, in the cockpit, one side has regular airplane controls while the other has controls similar to those of the shuttle. Pilots practice as many as a thousand approach and landings under different lighting and weather conditions to get the "feel" for the shuttle, again to train their reflexes. Long hours are spent also practicing approach and landings at the shuttle mission simulator (SMS), where there is a movable cockpit called "the motion base" simulator. A second cockpit at the SMS is called "the fixed base" simulator, which allows practice of on-orbit mission phases. Both are exact duplicates of the shuttle. The pilots spend their final 120 hours in these simulators with members of their specifically assigned crews, going through the flight from start to finish.

Pilots make mistakes in these trainers. Instructors can remember times when astronauts killed themselves several times in one day by making wrong moves in the simulator. Although there are no formal tests for these training sessions, no ranking according to performance as in flight school, astronauts learn quickly from their embarrassment.

Pilots sometimes take on specialized piloting technical assignments. For instance, before the flight of STS-1, pilot-astronaut Ron Grabe and many others were assigned to studying problems that could arise during ascent or entry. "This involved working in the simulators quite alot," said Grabe, "probing all questions of operation and engineering, trying to come up for the optimum procedures for emergency situations."

Mission specialist training involves less flying, of course, but a wider variety of assignments. Many are called upon at one time or another to work with companies contracted to provide special equipment for the shuttle. The astronaut can test prototype equipment and can critique impractical aspects of an equipment's design and operation. "Most people who work on this equipment aren't really aware of what it is like to work in space," said former astronaut Alan Bean. "An astronaut can come in and say, 'Well, this is too complicated and too time-consuming to use as is in space.' Then he or she can make suggestions on how to make it useful in space."

"In the future," said Frank Hughes, "astronauts will be working on a regular basis with specific corporations — Hughes, Ford,

McDonnell, and so on — involved in ongoing shuttle projects. We had thought about specializing people in the launch of certain special satellites, like the TDRS (Tracking and Data Relay Satellite) or those using an IUS [a special kind of rocket stage which will boost a satellite into geosynchronous orbit]. But we haven't yet seen the need for that kind of specialization."

A useful technical assignment for learning about all aspects of the shuttle is the Flight Data File or FDF. The flight data file is, essentially, the owner's manual of the shuttle. It weighs about 70 pounds, consists of more than twenty different books plus maps and charts and 100 cue cards. It contains a complete listing of all conceivable flight procedures — step-by-step directions for what to do if any of the thousands of systems malfunctions. The FDF is always being updated, in order to update changes in shuttle software and hardware, to allow for new payloads, and so forth.

One of the long assignments for which both pilots and mission specialists can qualify is that of capsule communicator or CAP-COM. Since the early days of the space program, astronauts have always communicated with one individual in Mission Control — another astronaut, whose sole purpose it is to relay messages to the spaceborne astronauts. He or she sits at the right hand of the flight director and listens to most discussions about specific problems that come up, conveying only what is necessary for the astronauts to know and radioing up the flight director's commands. Naturally, this cuts down on radio chatter since it is only through the CAPCOM that Mission Control personnel can convey messages. Like SAIL, the CAPCOM position gives the astronaut a comprehensive under-standing of the shuttle systems and various problems that can occur. The one exception to the CAPCOM rule is made during Spacelab flights when scientists closely involved with experiments aboard are allowed to speak directly to the astronauts.

Specialized training for astronauts who are being groomed for special assignments include spacesuit training for spacewalking or EVA, something that is carried out in a giant swimming pool called the WETF (weightless environment training facility). The pool is 25 feet deep everywhere (no shallow end), and the suited astronaut is lowered into it by way of a crane. According to Story Musgrave, who trained extensively in the WETF: "The divers detach you from the crane and add lead weights to the suit to get you in a state of neutral buoyancy. You're not truly weightless because you can still feel pressure points in the suit. And you have viscosity in the tank. It's hard work pushing water aside, whereas in space, there are no pressure points and there's no viscosity. In space, there's a lot more mass moving because you have on a 480-pound backpack, so you move slowly. In the water, an umbilical cord is used for life support, so the divers have to always make sure you don't torque up." There are additional safety divers on the surface, documentation divers who photograph the session, and other divers involved with special equipment, so the astronaut is surrounded by help all the time. In space, only the astronaut and his/her partner are there.

Every EVA session is carefully "choreographed," said Musgrave, and all "movements are carefully verified." Special equipment is lowered as well, which the astronaut will have to get accustomed to using or repairing in a weightless environment. Some tools fail to work properly, have to be redesigned, and so on, after WETF testing: "We had a winch problem during our spacewalk," said Don Peterson, "which had also malfunctioned in the WETF. But the contractor believed it malfunctioned because it was underwater and not in space. It turned out to be a mechanical difficulty and the hardware really did have to be modified."

According to Frank Hughes: "We hope to have a cadre of about 15 EVA experts who are specialized in spacewalk procedures be-cause we are constrained by the numbers, sizes, and expense of the spacesuits. The selection of this group will depend on their technical and training experience, possibly."

Another possible area of specialization within the astronaut mis-sion specialist category may be in the use of the RMS (the remote manipulator system or "arm" that is used to hoist satellites in and out of the cargo bay). Mission specialists learning to use the RMS spend a great deal of time in the RMS trainer at the Johnson Space Center, a simulator that has a duplicate arm, the exact controls, and large helium balloons to practice with. According to both Sally Ride and John Fabian who operated the arm during STS-7, the simulator was an excellent practice device with very high fidelity operating capabilities.

On rare occasions, astronauts are assigned technical tasks in which they can use their area of specialization. For instance, Rhea

Seddon, a former surgeon, was assigned briefly to stocking the shuttle medical kit. Jeff Hoffman, an astronomer, was assigned to work on the large but delicate space telescope. William Thornton and Norman Thagard, both physicians, were instructed to conduct medical experiments on their flights exclusively.

Nonetheless, Hughes argues that the cases of Thagard and Thornton — astronauts practicing their scientific specialties — are "exceptional, because of the need to find some remedy for space adaptation syndrome. As a rule, we're going to try to keep the astronauts generalists. They came here with the idea of flying spaceships as a career. Some may feel as if they lack challenge and career development after several flights and may burn out. Others with scientific specialties will do that on the side if they have time. But it's unlikely that we'll have an astronaut corps made up of scientific specialists any time soon."

Mission specialists sometimes take on a number of technical tasks at a time. For instance, early in her career, Anna Fisher helped Hamilton-Standard, the manufacturer of the spacesuit, develop a comfortable suit for women, whose upper body strength was not as great as men's. She also tested a shuttle tile repair kit in which epoxy was sprayed into the space of a lost or broken thermal tile, without the unfortunate effect of propelling her backward. (The repair kit was shelved after the first and second flights of Columbia proved it to be unnecessary.) She helped Martin-Marietta in Denver test and develop the manned maneuvering unit (MMU), the rocket-powered backpack that allows a spacesuited astronaut to propel himself/herself around the shuttle and beyond, without use of handrails and tethers. Later, she worked at SAIL.

Sally Ride, on the other hand, specialized very early in the development of the RMS and showed uncanny skill at using it, possibly because, according to one NASA technician, "she was a champion tennis player and had such extraordinary hand-to-eye coordination." Later she became CAPCOM, in which she mastered Mission Control operations with such professionalism that she impressed many people — not the least of whom was Robert Crippen, who asked her to be named to his STS-7 flight, one that would demand both her expertise at the RMS and her knowledge of shuttle systems in general.

There are numerous technical assignments, and new astronauts are out of town roughly one third of the year, since they provide technical aid to industries involved in the space program all over the country. An astronaut may have to fulfill a speaking engagement anywhere in the world. "A ten- or twelve-hour day is typical," said Bill McClure of the Astronaut Scheduling Office.

Nobody minds. It's just part of being an astronaut. "When you like your work," said Bonnie Dunbar, "you don't mind putting in the long hours. Sometimes, if you're on a strange shift, you might not be able to do everyday things, like shopping or running errands, at conventional times."

During these technical assignments, the astronauts are always hopeful for an early flight assignment. However, flight assignments are the province of George Abbey, head of flight operations at the Johnson Space Center. And nobody can second-guess him. "Flight assignments are a mysterious process," said one astronaut, a process upon which astronauts are unwilling to speculate or even reflect.

Crew-cutted, stocky Mr. Abbey, in his characteristic form of understatement, says about crew flight assignments, "Oh, I get involved in it." He claims that "a lot of people make recommendations for crew selections on particular flights," not the least of whom is the flight's commander. Robert Crippen has requested—and gotten —Sally Ride assigned to his crews twice. According to Abbey: "We look at the requirements for any particular mission, look at who's available and best qualified for that mission—what we have to do on that mission." Although the pattern has been to fly a veteran pilot-astronaut with a rookie pilot-astronaut, "there's no reason why we couldn't depart from that pattern if we needed to. Certainly it's desirable, from the standpoint of training, to fly a veteran with a rookie, but I would envision the point where it might be hard to do that kind of thing with the frequency of flights and the training requirements of the missions. We might not be able to do that all the time." Abbey is perhaps the keenest on keeping astronauts as "generalist" and "unspecialized" as possible since maximum flexibility of crew assignments is achieved that way. "To be useful to us, they have to be generalists. If they get too specialized in one area, we can only use them in that one function. Look at it in the context of a particular mission and hopefully we have many people who can satisfy the mission requirements."

Abbey has been perhaps the greatest power in forcing individual astronauts to expand their capabilities to the maximum. For instance, Judy Resnik had long experience in the use of the shuttle arm. However, her initial flight assignment on STS-12 did not call for that specialty but rather required a great deal of photographic work. When the flight was manifested differently, the arm work came back and Resnik had a chance to exercise her specialty. Originally, however, she was not picked for this capability but, perhaps, for her remarkable professional flexibility, her ability to master tasks quickly.

However, Abbey is not above naming a specialist if the job requires one. "If there is a telescope on board," he said, "we have a number of trained observers. But that depends on how specialized the telescope is. If it requires an astronomer, then we have astronomers."

Because Abbey wields such tremendous power over their professional lives, the astronauts are all eager to do as he bids, to take on the disagreeable tasks wholeheartedly, to show the willingness to sacrifice and the ability to dedicate themselves wholly to the demands of the space program. Abbey sidelines the astronauts who show signs of balkiness, lack of competence, and unwillingness to cooperate. To be sidelined is to be placed in professional limbo. One must stay with a technical assignment indefinitely while one's colleagues are, one after the other, assigned their places in space. It is a professional situation the astronauts wisely avoid — like plague.

Once an astronaut is given a flight asignment, general training ends, and he or she devotes himself or herself to that flight alone: the manifest of experiments and payloads aboard, the special requirements of the orbit, and so on. All training three to six months before launch is devoted exclusively to the specific tasks and the specific flight. Three months before flight, the schedule becomes very hectic, often adding up to 80 or 100 work hours per week. Much of this training time is passed in the full-scale simulators in the SMS (Shuttle Mission Simulators).

To enter one of these simulators is to come as close to real life on the shuttle as one can get on Earth. Everything is there: the "swizzle stick" poles that the astronauts use to reach switches that are impossible to reach when they are suited up for launch; the visual displays of Earth outside the window, "not as pretty as it really is and not in as great detail," according to all the astronauts; the live displays and switches; the hum of the instruments and fans; and the feel of the narrow pilots' seats. Once while I was on assignment in the SMS with a *Science Digest* photographer, Roger Ressmeyer, the cabin was plunged into absolute darkness. All the instruments went black except for the visuals outside the window, which still made us feel as if we were orbiting the Earth. Even though neither of us was afraid of the dark, we both felt momentarily uncomfortable. The interior lights and visual displays soon came on again, but I can remember feeling a distinct sense of relief. For some reason, the lights reminded me of Christmas. Roger and I came away with a new appreciation for technology, for how much one's life depends on it in space.

The motion base is used primarily to train astronauts for launch and landing, since it is set up to jiggle, swing around, and simulate those physical perceptions of launch and landing, including the noise. Like the fixed base, the motion base simulator (MBS) has computer-simulated visuals projected out the windows. One training director told us that the trainer is so high fidelity that the astronauts can get very uncomfortable in simulated emergencies, quite forgetful of the fact that their lives are in no real danger while they are in the simulator.

All the astronauts have remarked that their training in the simulators was very close to their real experiences in space, as if "we had already lived through those days many times." Several agreed with John Fabian who said: "Actually, doing a simulation is in some respects harder, because the actual flight did not have most of the problems we had to overcome in training."

In the past, astronauts were trained by a large number of personnel with a wide range of specialties. However, flight operations upper management decided it might be more efficient to assign each crew a small set of specialized trainers who work with them exclusively. "That group will be in the simulator every time the astronaut is," said Hughes. "That way, the astronauts get to know the people personally. The trainers can test them and observe them and can assess their training needs. They can suggest that an astronaut needs more work in a certain area. The training team would be led by one manager who is in charge of that flight and a subgroup of experts who are all knowledgeable in aspects of that specific flight. It's difficult for a young GS-7 training person to tell a veteran like John Young what to do, of course, but if the GS-7 works with John everyday, the barriers break down and John would listen to what that person might have to say."

The current training procedure includes a list of tasks the astronauts must know. A record is kept of which skills have been mastered and which have not. The training is intense and personalized, said Hughes. "An astronaut can say, 'I need more work on this,' and our training is flexible enough now that he or she can do that and forego skills he or she is well trained in. Also we only have a couple simulators and we can maximize use of them in this way."

When simulations are done with the three shifts of flight controllers at Mission Control, the simulation is called an "integrated sim" and can last a few hours to a few days. The crew is essentially put through practice runs of the actual tasks aboard the flight, and their proficiency and that of the flight controllers is tested by the addition of certain malfunctions, usually ones that involve more than one system and are difficult to fix.

The integrated simulations are run by training staffs organized by what one astronaut called, "especially perverse people," whose diabolical imaginings seem inexhaustible. The "sim-sups" (short for simulation supervisors) as they're known in the vernacular, are sometimes baptized with nicknames most fitting to their dispositions, like Dr. Doom.

Integrated simulations are as close to real-life space flight as an astronaut can get on Earth. In some simulations, two auxiliary power units might go down, propellant might be leaking, and two of the five computers on board could go down . . . all at the same time. Sometimes, whole missions were aborted during sims because certain problems could not be solved.

Almost all the astronauts agree that the simulations are more difficult than the flights. Most people involved agree that, if anything, everybody is somewhat "overtrained."

John Young and Robert Crippen, who flew the first shuttle into space, had 1500 hours of formal mission-specific training. In future, total astronaut training time will be more like 350, and veterans may require as little as 115 hours to prepare for a flight. "Many things have occurred to bring down the amount of training time," said Hughes. "For one thing, a lot of training hours went away once we understood the shuttle better. Young and Crippen were trained in exotic aerodynamic problems that theoretically might have happened but in fact never did. Generally, we're cutting down on training for exotic malfunctions and train now for what the astronauts really need. For instance, we only rarely sim for an unlikely scenario. (One recent one I can think of involved flying the Orbiter like a catcher's mitt to get a stranded or unconscious spacewalker back into the payload bay.) Also we're cutting down some training time by doing things like standardizing stowage. For instance, we're putting all the underwear in compartment A-3 and so on. Endless changing of stowage takes up astronaut time."

Stowage schemes are learned in an exact duplicate of the shuttle middeck called the "one-g trainer." Astronauts are walked through the one-g trainer in order to become familiar with the food system, the toilet, some middeck experiments, and the airlock.

Mission specialists and payload specialists assigned to Spacelab missions have an exact duplicate of the Spacelab in which to do practice runs of experiments.

Finally the astronauts go through "vehicle familiarization," that is, getting to know, touch, feel the actual Orbiter, in two stages. First, the astronauts are flown to the Kennedy Space Center in Florida before the shuttle is mated to the propellant tanks "in order for them to get to know everything about it — even the payload bay — every sharp edge, every bolt," said Hughes. "It is especially important for people preparing for a spacewalk to know where every sharp edge is."

The second part of familiarization comes a few days before launch, in which launch abort sequences and launch scenarios are practiced with the Kennedy Space Center launch control team. The astronauts practice emergency procedures, egress problems — the kinds of problems that can occur within the first minute of the flight.

Training has been quite successful thus far. Some of the first astronauts to fly the shuttle complained that they were perhaps overtrained in exotic malfunctions, but this was remedied after a few flights. The only failure mode for which they had not simmed occurred on STS-3, in which, according to Hughes, "Jack Lousma made the best of a terrible situation. He landed in a bad wind and had to take over from the autopilot much too close to the ground. There were two software problems: one caused the speedbrake to close when it should not have, so the Orbiter was coming in too fast; the other faded out the pilot's input, that is, filters didn't permit the shuttle to respond to the commander's rapid hand movements. As he was coming in, he knew he wasn't supposed to allow the nose to touch the ground until he was going slower. But Jack needed a rapid movement to keep the nose up. Well, the way that worked was that he pulled back to keep the nose up and nothing happened, so he pulled back some more and the nose reared up, because the inputs from the first motion were just coming in. So he just stuck the gear down again and the nose came down. If he hadn't been the reactive kind of guy he was, the Orbiter could have become airborne again or flipped over. It would not have been a good day. Fortunately, Jack was overtrained and had spent a lot of time in the simulators."

Training managers have thought about making the shuttle Orbiter itself a trainer for some specialized future missions. Said Hughes, "The astronauts may simulate OTV [orbital transfer vehicle — little spaceships that will be used to fetch and carry satellites from the station to geosynchronous orbit or back] activities in space itself — using desk-sized arcade trainers to train themselves. To keep proficiency up for OTV, we can place OTV trainers on board the station and have visual sims of problems. Take your training with you."

Of course, with a stock of veterans flying, rookies will learn not only from trainers but also from experienced pilots — the do's and don'ts of the beast, the subtle peculiarities that are difficult to train for.

Future training may differ in other respects, too. Now, flights and flight simulations are rigidly structured around a central crew activity plan (CAP). In future the CAP may become more flexible, and scheduled experiments will not be so tightly timed that the astronauts feel rushed all the time. "Instead, the commander would have a list of tasks that must be done for reasons of orbital position, and a list that should be done whenever it's convenient," said Hughes.

"We would leave it to the commander to decide when these other tasks should be done. In this way, the commander would act like a commander on a naval vessel. A good commander can maximize efficiency of the crew. People on the ground sometimes have misapprehensions on how long it takes to do certain tasks in space — it's not uncommon. So why not stack up those jobs and have the commander assign available crewmembers to certain tasks? When an astronaut says 'I'm done, now what?' the commander can say, 'Do that.' Or if someone feels sick, the commander can say, 'Go up on the flight deck and take pictures.' In space, they already take some liberty with flight plans because they have to. This goes on the philosophy that the guy on the spot is the only one cognizant of what CAN be done. Overplanning, hyperplanning is unnecessary."

Future space station training might again involve some overtraining in the early years, the "ready for Armageddon" mode that characterized the early space shuttle program. "Scientists working aboard a station will probably only get payload specialist training," said Hughes. "Just about 100 – 120 hours — how to eat, sleep, get out of the way in an emergency. The pilot and the commander of the space station will have much, much more training in order to handle the emergencies."

Most astronauts feel that the public has serious misconceptions about what the astronaut job really involves.

"Operating a spacecraft is a real engineering skill," said Grabe. "It's not just fun and games and public relations."

"People think we're tossed into centrifuges, dunked into water and thrown out of planes in this work," said Sally Ride. "But it's not always exciting. We sit behind desks and go to meetings mostly."

"If you're an astronaut, your prime interest in life has to be doing space," said Alan Bean. "The real hard dog work is more prominent than the glamour of flying. It's easy to get discouraged."

"I fully expect some astronauts will get burned out," said Hughes.

However, the astronauts of 1978 and 1980 are still far from that. Their enthusiasm and dedication spill over into their public appearances — the numerous speeches they have to make, the television and radio spots they have to do. The attitude of most is that of Bonnie Dunbar: "Most astronauts will be in this business for as long as they can fly. What other job can possibly compare to this?"

Fig. 1 Rockets with seven million pounds of thrust blast the STS-4 astronauts into orbit, a thrill Sally Ride later described as "an E ticket at Disneyland."   (NASA)

Fig. 2 The shuttle, as seen by the free floating shuttle pallet satellite (SPAS), racing swiftly over the Pacific at approximately 17,000 miles per hour. (NASA)

Fig. 3 Professionalism and humor bind many astronaut crews together. Here, astronauts William Lenoir (upper left), Robert Overmyer (upper right), Vance Brand (lower left), and Joseph Allen (lower right) clown during a light moment aboard STS-5. (NASA)

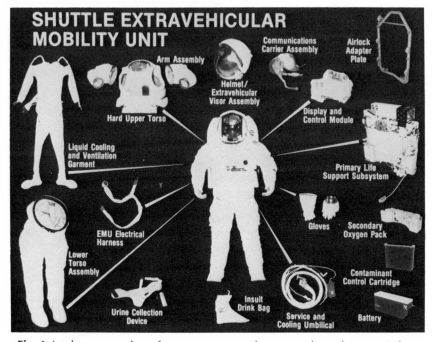

Fig. 4 It takes over an hour for an astronaut to don a complicated spacesuit for a trip outside the shuttle cabin. (NASA)

Fig. 5 "You really feel like you're trucking on down the highway," said mission specialist Story Musgrave of spacewalking. Musgrave and fellow-spacewalker Don Peterson along with the shuttle and their tools careened through space at a rate of 17,000 miles per hour during the STS-6 flight. (NASA)

Fig. 6 Musgrave slowly makes his way to the shuttle's aft end along built-in handrails. (NASA)

Fig. 7 During the 41-B flight mission specialists Bruce McCandless (left) and Robert Stewart (right) work in the shuttle bay, not far from the open airlock hatch.
(NASA)

Fig. 8 McCandless slowly backs away from the shuttle, reflected in his helmet, during the first test of the rocket-powered backpack called the manned maneuvering unit (MMU). (NASA)

Fig. 9 McCandless (shown here) became the first free-flying human satellites to orbit the Earth. With the help of the MMU, he experienced completely untethered human spaceflight for the first time. (NASA)

Fig. 10 McCandless tests a "cherry picker" platform, attached to the shuttle arm. Such a platform will be used in future satellite repairs. (NASA)

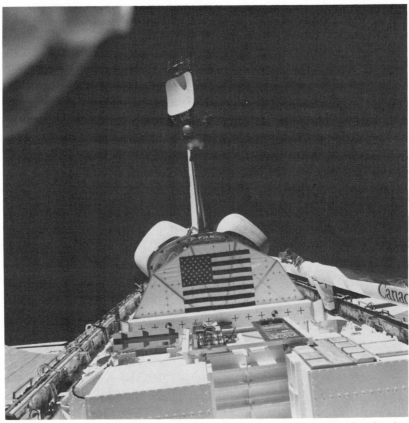

Fig. 11 The main work of the astronauts is launching satellites from the shuttle's payload bay. Here, the STS-8 astronauts launch the *INSAT* communications satellite for India.
(NASA)

Fig. 12 A future activity of astronauts may be the construction and alignment of large space structures. (NASA)

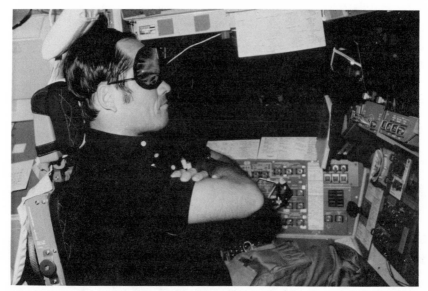

Fig. 13 STS-7 commander Robert Crippen sleeping in the "imperial" position — strapped into his seat with arms folded so that they don't float. (NASA)

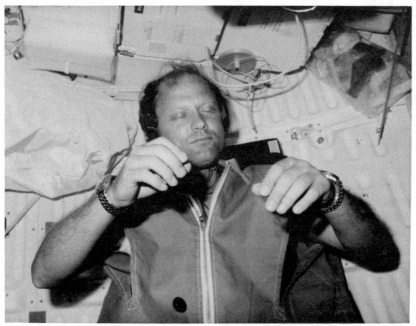

Fig. 14 Arms float independently in front of him as STS-7 mission specialist Norman Thagard sleeps soundly. (NASA)

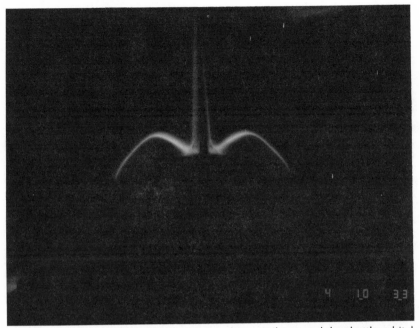

Fig. 15 A space version of St. Elmo's fire glows eerily around the shuttle orbital maneuvering system (OMS) pods. (NASA)

Fig. 16 A ghostly shuttle Orbiter materializes just beyond the runway footlights during the night landing of STS-8. (NASA)

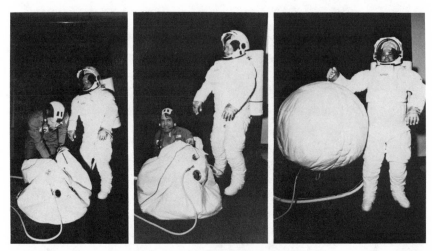

Fig. 17 Astronaut climbing into a "rescue sphere," which is currently used as a test of claustrophobia. (NASA)

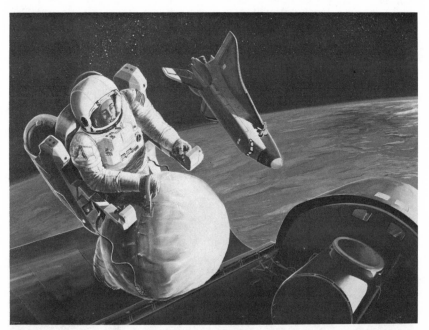

Fig. 18 An artist's conception of the rescue sphere being used to transport people from a crippled shuttle to a rescue vehicle. (NASA)

Fig. 19 Husband and wife astronauts Robert Gibson and Rhea Seddon examine an engine just before a routine session of T-38 jet training. (NASA)

Fig. 20 Astronauts routinely take on specialized technical assignments to test special equipment. Here, Anna Fisher tests a shuttle tile repair kit. (NASA)

Fig. 21   Fisher floats freely briefly in a KC-135 aircraft in preparation for spaceflight weightlessness. (NASA)

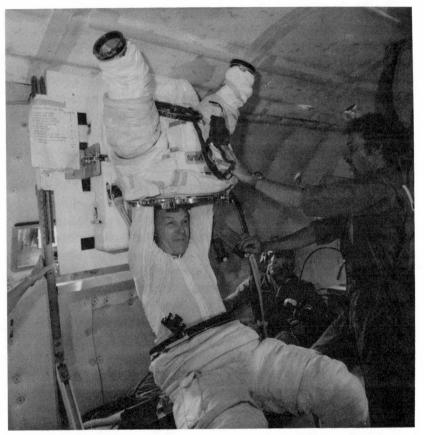

Fig. 22 Astronaut Story Musgrave practicing suiting up in a KC-135. The only way to get into the hard-shell torso section is to do a "breast stroke" maneuver.

(NASA)

Fig. 23 The Shuttle Training Aircraft (STA) is altered aerodynamically to fly like the shuttle Orbiter. Considered a very important training vehicle, the STA is used to teach pilots to land in all kinds of weather. (NASA)

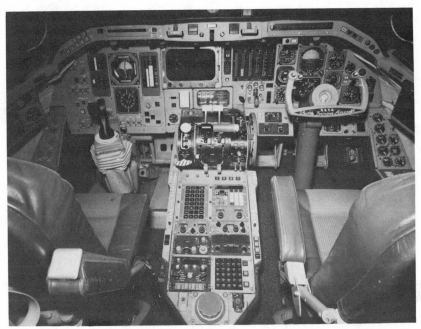

Fig. 24 The cockpit of the STA specially equipped with shuttle controls on the left and ordinary aircraft controls on the right. (NASA)

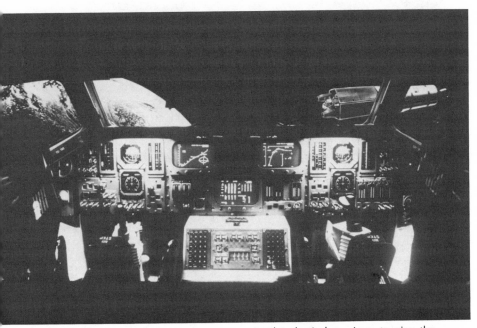

Fig. 25 Shuttle simulators use computer-simulated window views to give the astronauts a sense of realism during the practice sessions. Here, a satellite targeted for repair or retrieval is projected oustide the windows of the motion-base simulator. (NASA)

Fig. 26 All of the STS-6 crewmembers take part in an "integrated" simulation in the fixed base simulator. Such integrated simulations can go on for days and are used to prepare astronauts and Mission Control for potential spaceflight problems.
(NASA)

Fig. 27 The giant swimming pool at the Johnson Space Center, called the WETF (Weightless Environment Test Facility), is used extensively to train spacewalkers and to test equipment in a neutral buoyancy environment, somewhat similar to space weightlessness.
(NASA)

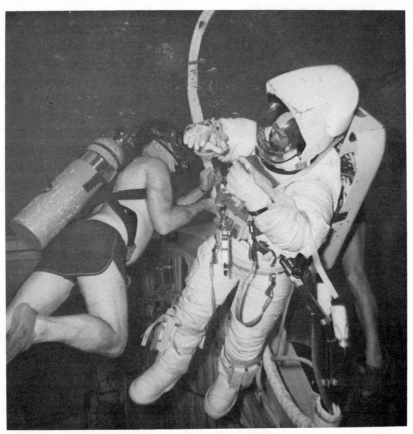

Fig. 28 Safety divers assist STS-1 backup pilot Richard Truly in the WETF during a training session. (NASA)

Fig. 29 A future spacewalker gets the feel of using the manned maneuvering unit at Martin-Marietta's Denver facility. (NASA)

Fig. 30 STS-7 mission specialist Sally Ride trains to use the robotlike shuttle "arm" or remote manipulator system (RMS) at the manipulator development facility at Johnson Space Center. (NASA)

4

Scientists, Ahoy

Payload specialists probably are representative of future spacemen in terms of their training.

— NASA Training Manual

There is so much we don't know about space. Earthbound materials change their very natures when they are put into a world in which weight, convection, and other "natural" phenomena are absent. They change even more profoundly when they are placed in the frigid vacuum outside the shuttle cabin. To see molten metals, fluids, and organic systems like plants and animals function in space is to see them with new eyes, to grasp their hidden potentials.

The pathway that scientists and engineers will take into space to study these new manifestations, to grasp the nature of these new potentials, and to apply them to future use will be that of the "payload specialist," a special category of scientist passenger.

Spacelab 1 was the first flight of such space people. Spacelab, a cooperative effort between the European Space Agency and NASA, flew a wide range of experiments from both American and European universities and research centers, 73 in all. Thirty-eight experiments involved materials science, that is, exploring the behavior of various materials in space, while the other 35 ranged over Earth resources, atmospheric, solar and space plasma physics, astronomy, and life sciences. The two payload specialists who joined the four NASA astronauts were Byron Lichtenberg, an MIT scientist, and Ulf Merbold, a West German scientist. Neither was selected by NASA, neither had to go through arduous astronaut training, neither had to give up scientific careers for the opportunity to fly in space. But both represented the type of activity that will dominate future space activity: the systematic exploration of space, from the behavior of the smallest gaseous particles to the measurement of the edges of the known universe. The whole pace of science will quicken immeasurably as a result of flying scientists' presence in space. "There will be a real avalanche of scientific ideas when the shuttle goes up frequently and the budget allows us to either build a space station or fly a large number of Spacelab missions," said Byron Lichtenberg. "Now it takes six or seven years to propose, design, and fly an experiment. In the late '80s it might take only six months."

Although being selected to fly on the shuttle as a payload specialist is still no easy feat, the manner of selection differs greatly from NASA's procedures for selecting career astronauts. The NASA selection is open to anyone who wishes to apply whereas payload specialists are nominated by the principal scientists who are flying their experiments on a particular flight. "Each scientist with an instrument on board Spacelab 1 nominated two people," said Dr. C. Richard "Rick" Chappell, chief mission scientist for Spacelab 1, who chaired the selection committee, "and he had to ask himself, 'What kind of person do I need to do an effective job running my instrument?'"

Byron Lichtenberg was completing a Ph.D. in biomedical engineering at MIT when the principal investigator for one of the medical experiments nominated him. Lichtenberg had wanted to be an astronaut since he was seven. During high school, he read science fiction, and by the time he entered college, he had begun to acquire the prerequisites for being an astronaut, which at that time involved being a military test pilot. He enrolled in Air Force ROTC, got an undergraduate degree in electrical engineering, went through pilot training, and flew F-4's during the Vietnam war, receiving two Distinguished Flying Crosses and 11 other air medals. He still flies A-10 fighter planes for the Massachusetts Air National Guard a weekend every month.

By the time he got out of Vietnam in 1973, the space program was in a slump, and he left the Air Force for graduate school at MIT. His master's degree was in mechanical engineering and his doctorate was in biomedical engineering. His work at MIT included building artificial limbs for a while.

In the mid-1970s, former astronaut Phillip Chapman, an electrical and aerospace engineer who left NASA before getting a chance to fly in space, spoke at MIT and rekindled Lichtenberg's dream of flying in space. An eloquent enthusiast for all space projects, Chapman encouraged Lichtenberg to visit NASA and speak with astronauts about areas of future space work. Astronaut Story Musgrave pointed him in the direction of studying space sickness and the vestibular system's adaptation to weightlessness. Once back at MIT, he "jumped horses" and shifted from studying artificial limbs in the biomedical area to working with MIT's aeronautics and astronautics department whose proposal won him his berth on the shuttle. "I had tried for an astronaut job just a few months before," said Lichtenberg, "and the principal investigator knew I was really interested in flying in space so he nominated me."

Forty nominations for payload specialist candidates came from the principal investigators. After submitting their resumés, references, and applications, they were given two interviews, one on general knowledge of science and the other on specific Spacelab experiments.

"The interviews are a lot like a Ph.D. oral exam in which the candidates could display their knowledge," said Chappell, who had a Ph.D. himself in space science and was chief of the solar-terrestrial physics division at Marshall Space Flight Center.

"In the first interview, they asked me things like: 'What is the lifetime of a red blood cell in the body? What is the concentration of atoms and molecules in outer space? What does the shape of the magnetic field of the Earth look like? How do you grow a crystal and why do you go into space to do that?'" said Lichtenberg.

After the first interview, the candidates were given the thorough NASA medical and psychological tests, including psychiatric interviews; various blood studies; hearing and vision tests; cardiovascular evaluation, including treadmill stress tests and resting electrocardiography; TB, hepatitis, and venereal disease tests; pregnancy tests for all females; and stool and complete urine analysis. "You can always tell who the potential payload specialists are when they're down for the medical exams," said one engineer at the Johnson Space Center. "They always carry little blue urine sample bags around."

After the battery of medical tests, Lichtenberg recalled being given "a 25-pound box of documents on the specifics of the Spacelab 1 experiments. We were told, 'You might want to look at these before the next interview,' and , like a good student, I read them all."

During a second interview, the finalists were grilled on the minute specifics of the mission. "They asked things like: 'What is the Bridgefort method of growing crystals, and draw a ray diagram of this instrument and tell us why it has an aspherical mirror rather than a spherical one,'" said Lichtenberg.

"The second interview tested both motivation and the ability to learn," said Chappell.

The extreme emphasis put on flying experience in the regular astronaut selections is absent as a criterion for payload specialist. "You're dealing with a group of scientists who are not enamored of people who fly planes, and they're picking someone who can run their experiments well," said Chappell.

Of the 40 candidates, only two American payload specialists, Lichtenberg and Michael Lampton, a research physicist from the space science laboratory at the University of California, Berkeley, were chosen, one to fly and one to serve as backup.

The European Space Agency, by a separate process, also selected a prime and a backup payload specialist, West German physicist Dr. Ulf Merbold and Dutch physicist Dr. Wubbo Ockels.

Because Spacelab 1 had a wide range of scientific experiments in many different fields, the Spacelab 1 selection committee sought an individual who was more of a generalist; however, this will not be typical of future Spacelab flights. "Many Spacelab flights will be wholly dedicated to one scientific specialty," said Chappell, "and will require very expert specialists in that one area. For instance, Spacelab 6 will be dedicated to plasma physics, so we will need people who are plasma physicists." The training in these specialized flights will be shorter since all the people will be experts and need only familiarize themselves with the shuttle equipment, something that will take a few months. Spacelab 1 training took five years in which, as Lichtenberg says, "I got the equivalent of five or six graduate degrees," familiarizing himself with equipment in the various scientific areas.

The primary difference between a mission specialist, a NASA astronaut who primarily performs scientific experiments, and a payload specialist is that "the payload specialist has some unique skills within the given discipline of a specific flight that the mission specialist doesn't," said Chappell. Also a payload specialist will probably fly only once or twice, whereas career astronauts may fly 20 or 30 times.

The scientific opportunity of such flights will be unparalleled. In Spacelab 1, for example, Lichtenberg and Merbold were able to analyze the data on the spot, judge whether an experiment was working properly, and fix it if it wasn't, a flexibility that ordinary shuttle astronauts don't usually have. Lichtenberg pointed out that on the medical experiments, "We can push ourselves more and do more invasive kinds of things without worrying about whether we'll be well enough to fly the shuttle home." He tried various ways to induce motion sickness, to test the adaptation of the nervous system, the senses, and the organs of balance to zero gravity. Blood was drawn three or four times in the mission, and electroshock therapy was used to test motor reflexes.

During Spacelab 1, Lichtenberg wore a metal accelerometer headband while he was whipped around in a floating chair. Cameras carefully recorded his eye movements. Astronaut mission specialist Bob Parker became drowsy when a rotating drum of dot patterns was placed over his head. Astronaut mission specialist Owen Garriot endured the punishing "drop and shock" test in which he was given a small shock as he dropped from a metal bar to test his reflexes.

Other experiments crowded the Spacelab schedule. In materials science, crystal growth and the behavior of liquids fascinated the scientists. "We started out with a large solid silicon rod that we melted and then resolidified," said Chappell. "We wanted to find out how large a liquid zone you could have to grow a crystal. This is the first time this method of growing a crystal—called the floating liquid zone technique — was used. It's very important to crystal growing in space. Such crystals would be used in imaging devices."

In atmospheric physics, Spacelab was used as an observation deck from which to look back at the upper atmosphere. Said Chappell: "From there you can measure the properties of the upper atmosphere and the chemistry of the atmosphere on a global scale, something you can't do easily with sounding rockets from Earth.

Also we did imaging of the Earth with a metric camera and a microwave instrument."

In astronomy, there were three telescopes, two in ultraviolet and one in x-rays. "These are used to study stellar sources above the atmosphere which absorbs ultraviolet and x-ray wavelengths. If you want to understand the births and deaths of stars, where you have a lot of x-ray emission," said Chappell, "you need to measure in these wavelengths."

In solar physics three instruments measured the energy output of the sun. "The sun's output varies," said Chappell, "and is responsible for large climatic changes. To study the output of the sun is important for future prediction of where the environment is going."

Spacelab studied the source of auroras in its plasma physics experiments. In another experiment, electron beams were shot into the atmosphere around the shuttle to understand how ionized plasma gas interacts with the beam.

The most exciting results came from a circadian rhythm experiment. All organisms on Earth possess circadian rhythms. Before Spacelab, it was not known whether these were generated by subtle cues from the rotation of the Earth or were innate to the organism. A quick-reproducing fungus showed the characteristic circadian rhythm bands in space, producing a "Eureka!" kind of discovery that hardly ever occurs in science.

During the postflight press conference, the Spacelab-1 crew praised the flight as an excellent model for future Spacelab flights. "We were able to perform interactive science," said Lichtenberg. "We were able to work with the principal investigators on the ground, get their baseline experiments completed, and then get creative when design problems or interesting results occurred. It really pays to have scientists in space."

After 10 days in the bathroom-sized Spacelab, working 18-hour days for 10 days, going without a bath in all that time, and sleeping in coffin-like compartments at night, the crew all agreed they got along very well.

The Spacelab-2 selection was a dedicated flight. Of 220 proposals for experiments, almost all involved with solar physics, only 13 experiments were accepted. It was decided that the payload specialists had to come from within the experimental teams themselves, experts all.

From the group of experts, 16 candidates were formally nominated. The principal investigators met and decided that there should be "no theoreticians," so six were immediately eliminated. The physical examinations, which were slightly more strict at that time than now, eliminated another. "You have to pass a class III Air Force physical once a year, but the original selection had some quirks in it," said Dianne Prinz, a Spacelab-2 backup specialist. "For instance, they measured every angle of each joint on the body, but nobody knew why. We were also tested for claustrophobia in the rescue sphere, which to me was like the world's smallest gym — there had been so many people in it, it didn't have too good an odor."

One scientist had "a diffuse brain wave on the electroencephalogram," indicating, perhaps, a tendency to epilepsy. Another candidate, a British member, dropped out voluntarily.

Of the eight left, each was supposed to give a ten-minute presentation to the team of thirteen principal investigators. On the basis of the presentation, the four finalists — Dr. John-David Bartoe, Dr. Loren Acton, Dr. Dianne Prinz, and Dr. George Simon — were picked. "I guess we had the gift of gab," said Prinz.

Loren Acton was a research scientist working at Lockheed Space Science Laboratory in Palo Alto, California. He was a coinvestigator on one of the solar experiments.

John-David Bartoe was a research physicist from the Naval Research Laboratory in Washington, D.C., who was coinvestigator on an experiment studying ultraviolet radiation from the sun.

Dianne Prinz was also a research physicist at the Naval Research Laboratory, a coinvestigator on a solar physics experiment, and a specialist in optical instrumentation design. She and Bartoe had both designed optics and flight software for the instruments aboard the Spacelab-2 experiments.

Dr. George Simon was chief of the solar research branch at the Air Force Geophysics Laboratory and a coinvestigator on an experiment.

According to Prinz, after selection, the payload specialists acted as liaisons between the experimenters and NASA, defining page displays as they evolved, developing the mission time line, and working up detailed ground command paths.

Spacelab-2 payload specialists and mission specialists will not work in the ESA-built special module, as the Spacelab-1 astronauts did, but will stay within the Orbiter itself, monitoring experiments mounted on "pallets" or equipment-laden platforms in the shuttle payload bay. An infrared astronomy experiment will map sources of infrared radiation in the sky, while another will determine the abundances of helium in the solar corona, and yet another will measure the shuttle's effects on the Earth's magnetosphere from a deployed subsatellite. The advantage to having expert solar physicists aboard, according to Prinz, is "that there are decisions made based on seeing something. The TDRS [Tracking and Data Relay Satellite] does not have enough bandwidth to help scientists on the ground make these judgments. There must be eyes on board that can comprehend complex feedback from experiments. It's not just a question of switches and dials. And you can't dream up a procedure for every malfunction which may occur. In space, it's absolutely crucial to have people on board who are intimately familiar with the equipment."

Bartoe and Acton were selected to fly aboard Spacelab-2 as the prime payload specialists. Prinz and Simon may get their chance to fly later, possibly aboard Sunlab, another dedicated solar physics mission.

Spacelab-3 is another dedicated mission devoted this time to materials processing, especially crystal growth and fluid physics experiments.

In 1976, NASA put out a "cattle call," as one principal investigator put it, for Spacelab-3 experiment proposals. "There were, I think, 400 or 500 proposals submitted at that time," said Taylor Wang, one of the principal investigators, "and mine was one of the 14 selected. It's so selective that no trial-and-error experiments were accepted. We were in a better position to compete because we had done a lot of the ground work already."

Wang, program manager for materials processing in space at the Jet Propulsion Laboratory (JPL) in Pasadena, California, is the principal investigator on Spacelab-3 experiments and three other spacebound experiments. His prominence in the field as an inventor and an aggressive researcher played some part in his being named a Spacelab-3 payload specialist. "As a principal investigator, you develop the experiment and the hardware. When NASA looked to its astronaut corps, it didn't have the capability to really run the experiments. So I said, I'd like to go. If you're a candidate, you can't vote on accepting yourself, and you're absent from the whole process. There are a lot of interviews, evaluations, physicals, psychological profiles to judge if you are going to go nuts up there and if you are going to disgrace the country. Then NASA says, 'You guys look alright,' and that's it."

Mary Helen Johnston, a back-up Spacelab-3 payload specialist, recalled that the selection was "not a process of elimination selection per se. They asked who would like to fly with the experiments and whittled down the number to seven finalists they interviewed. After physicals, the number went down to four."

Wang's coinvestigator, Eugene Trinh, was selected, along with Dr. Lodewijk van den Berg, an international authority on crystal growth on Earth and in space.

The odd thing about the Spacelab-3 payload specialists was that only one, Mary Helen Johnston, was a native-born American. The other three were naturalized Americans: Taylor Wang was born in Shanghai, China, Eugene Trinh in Saigon, Vietnam; and van den Berg in Sluiskil, the Netherlands.

After working as a chemical engineer for nearly ten years, van den Berg went to graduate school at the University of Delaware from 1971 to 1975, receiving an M.S. and Ph.D. in applied science. Afterward, he worked for EG&G Corporation in Goleta, California, as a senior scientist responsible for a crystal-growing facility. An authority on vapor growth techniques, he designed and tested the Spacelab-3 vapor crystal growth system.

Eugene Trinh came to the United States for its educational opportunities. He received a B.S. in engineering from Columbia University, an M.S. in applied physics from Yale, two years later, and an M.Ph. and a Ph.D. in applied physics in the four years following the M.S. In 1979, after completing his brilliant academic career, he went to work as a research physicist at NASA's Jet Propulsion Laboratory for Taylor Wang. There he studied fluid mechanics and acoustics, frequently flying with his experiments on the KC-135. Like Wang, he holds patents in his area.

Most interesting of all, however, is Taylor Wang's circuitous route to space. Wang's family had fled to Taiwan to escape the communists when he was ten. "My father was a businessman who worked for a shipping company," said Wang. "He was always disappointed that I didn't become a businessman. Maybe someday I'll fulfill his dream." He came to the United States for study at UCLA, where he received a B.S., M.S., and Ph.D. in physics. After his graduate work, he joined JPL in 1972 as a senior scientist, specializing in containerless processing technology and the dynamics of fluids in space.

"A whole bunch of coincidences led me into space work," said Wang. "My background has always been in fluid mechanics and acoustical mechanics. I first started at JPL in acoustic levitation processes and then drifted into materials processing, which led me into the Spacelab-3 experiments. When you come to JPL, you aim to do experiments in space, but nobody says, 'I'm planning to go to space,' because that opportunity had never presented itself to anybody at that time. When the opportunity arose to fly with the experiment, of course, you take advantage of it."

However, once the opportunity presented itself, Wang did not hesitate to grasp it. "Space is the only place in which you can have extended zero gravity. One type of experiment that can't be done anywhere else is called containerless — experiments on liquids without using containers. Containers affect and alter what you're looking for in liquids. In fact they dominate what you're looking for. When you take containers away in space, the liquid becomes a droplet, and if it's molten, it becomes a molten sphere. What is the physics that governs that? We're doing very basic physics in order to understand drop dynamics."

The future applications of Wang's work may be the basis of industries his father never dreamed of. Wang said: "With containerless processing, there are whole sets of materials we couldn't do because we had to use a container to process. Novel glasses, spherical shells, other admissible kinds of material are all possible."

This ongoing research will lead to many flights of this type in the future. According to Wang, he will fly this and maybe one subsequent mission and then Trinh may fly afterward on later missions.

As a prominent project manager, Wang still believes that the Spacelab-3 flight will have an impact on him professionally. "For the first time we'll be able to do quantitative experiments and definitive experiments. Whenever you do definitive experiments, it's always important professionally."

In the future, he would love to work on a space station. But, he cautions, "You don't plan on that, but if it comes along, it's just great. If the opportunity presents itself, it would be great. My primary interest is to do experiments — it's the sort of thing you have to do to do experiments. I've been flying KC-135 a long time, and I'll do anything necessary for the experiments. During a three-month stay on a space station, I'd certainly be able to get a lot of science done."

Although Mary Helen Johnston did not have to brave communist invasions, foreign academics, and U.S. immigration like her colleagues, she did have to overcome prejudice and sexism. Like Byron Lichtenberg, Johnston had always wanted to be in the space program, "ever since Shepard's flight," she said.

In college, one of her professors had been John Lewellyn, a Ph.D. chemist-astronaut who had withdrawn from the astronaut corps a year after his selection. "Because of him," said Johnston, "I became aware that NASA was looking for scientists. I realized that it was possible to be a scientist and fly."

But there was prejudice to overcome. "I was the first woman to graduate from Florida State University in engineering and went on to the University of Florida for graduate work. It was traumatic for them. The first day I showed up, the dean told me he was sure I'd be in home economics in six months. By the time I graduated, he was one of my best supporters. I guess I overcame prejudice by not taking any of his male students and marrying them and getting them out of engineering. I heard that a couple women had gone into engineering, caught themselves engineers, and then dropped out. I minded my own business. The only time I had to fight was when I co-oped at Marshall Space Flight Center. Women students were not allowed to live off campus if they were under 21. That's hard to do when your work is in Huntsville. It's so unreasonable sometimes. But by the time I got to that problem the dean of engineering helped me. That was the only way to handle it at that time: to get some people within the bureaucracy to help."

But if she had to do it again, she'd do it a bit differently. "Looking back over a lot of years, things like that don't make me nearly as mad. I don't let it eat at me as I used to. I'm low key and very persistent, so I endured. It may take a little longer but you've got to be patient. Prejudice can go on for years and years and can have physical and psychological repercussions, if a woman doesn't look out."

She had co-oped at Marshall for six years before becoming a full-time employee in 1968. By 1973 she had a master's and Ph.D. in metallurgy, continuing to work on a variety of materials processing problems for Marshall. She believed materials processing was one of the main important reasons to go into space, although at the time she chose her major, materials processing in space did not yet exist.

In 1974, she participated in an all-woman crew of experimenters in a five-day simulation of a Spacelab mission. "It was a concept verification test, looking at what it would take to have a Spacelab," said Johnston. "They put together a dedicated payload at Marshall because they needed to know how difficult it would be to handle materials processing experiments in space — they draw out a lot of power and put out a lot of heat."

Like Lichtenberg, she had tried to get into the astronaut corps but hadn't made it, although her long experience in her specialty did eventually lead to space.

On Spacelab, though, wide-ranging experiments with exciting potential benefits will be hers to conduct. Said Johnston: "I see at least two practical materials coming out of this. One is nuclear radiation detector material that will result in regular production of it. There are very good and early uses of it, like in the medical field — detecting radioactive tracers in the body, mapping the body. And I hope for more metallurgical research on Spacelab 3. We can learn more about the limitations, the processes involved in metals when the metals are cooled. This is very hard to study on Earth. Furnaces being developed now can go to high enough temperatures."

Like Wang, she would love to work on a space station lab and for similar reasons. "Three to six months in space would be good because with most real research, you have to go through a certain amount of trial and error — make mistakes and try something again. And if we're really going to find the novel and unique things in space, that's what we're going to have to do. I really believe payload specialists will become very prominent in the future. Industry and science will need them."

Out of many Spacelab flights, "superalloys" may be born. "These will be produced in small quantities," she said, "and used in something fairly small, like a turbine blade of an engine. If you can improve its performance by 10 percent by using superalloys in crucial places, it will improve fuel efficiency and avoid breakdowns — save millions of dollars in repairs."

Spacelab-4 slots called for biomedical experts to do animal experiments, since the flight is a dedicated biomedical one. The four named were Dr. Francis Gaffney, a cardiologist at the University of Texas Health Science Center in Dallas; Bill Williams, a Ph.D. animal physiologist of the Environmental Protection Agency in Corvallis, Oregon; Robert Phillips, a physiologist, of the department of physiology at Colorado State University and Dr. Millie Fulford, a Ph.D. biochemist, of the San Francisco Veterans Hospital.

Millie Fulford, when she was Millie Wylie, in 1977, had applied to the astronaut program along with Mary Helen Johnston of Spacelab 3 and Dr. Patricia Cowings, who has experiments aboard Spacelab 4 in biofeedback techniques of controlling space sickness. For Millie Hughes Wylie Fulford, persistence paid. She never gave up trying to get in to space. As a child she dreamed of being an astronaut and jumped out of bed early in the morning to watch space cadet "Rocky Jones" on T.V. Later she was a "Star Trek" and "Star Wars" fanatic. When she first learned of the astronaut job in a scientific journal, she thought, "I'm too old" (she was 31 at the time) but then listened to the little voice within that screamed, "Apply!"

Although she didn't make the astronaut corps, like Cowings, Johnston, and so many other talented women who may find their way into the ranks of the payload specialist, she did continue her work in biochemistry, which led to the space job. In an interview in *Newsday* on October 17, 1977, before she had found out that she had been rejected for the astronaut job, Fulford was quoted as saying: "We all have dreams to follow. I want to be able to reach the age of 75 or 85 and look back on my life and be happy with it. That means going after the things that are desirable and at least trying for them. I don't ever want to say I gave up without trying." The important thing in her case is that she kept trying.

The same module used for Spacelabs 1 and 3 will be outfitted with life science experiments for Spacelab 4. The objective is to study the safety, well-being, and productivity of humans in space.

Fluid shift, cardiovascular deconditioning, lung function, blood chemistry, protein metabolism, mineral loss and decalcification, and space sickness will all be tested. The astronauts, four squirrel monkeys, and 48 rats will all be test subjects. Beyond that the effects of weightlessness on oak seedlings, wheat seedlings, and amphibian eggs fertilized in space will all be investigated. Dedicated to life sciences as it is, Spacelab 4 promises to be interesting. A second flight of these experiments, called Spacelab 4B, may follow a year later, in order to verify the results of the first flight and refine a few of the experiments for modification and reflight.

Other missions involving payload specialists include Astro-1 to 3 to be launched in March 1986, November 1986, and July 1987. Using three ultraviolet astronomical telescopes, payload specialists will view the stars through video terminals. Three payload specialists were picked from a field of nine experts in this type of equipment. Two payload specialists will fly on each mission, and each payload specialist will be able to fly a total of two times. Samuel T. Durrance from the Johns Hopkins University, which built one of the telescopes, the Hopkins Ultraviolet Telescope; Kenneth H. Norsieck is from the University of Wisconsin, expert in the Wisconsin Ultraviolet Photopolarimetry Experiment; Ronald Parise from Goddard Space Flight Center, responsible for building the ultraviolet Imaging Telescope. They will study Halley's Comet and a number of other celestial bodies. "Spacelab officials will face a new challenge in meeting the turnaround time for the second and third Astro flights," said Micheal Sander.

EOM-1 (Environmental Observation Mission) is a mission designed to follow up on Spacelab-1 experiments. It has three payload specialists: Spacelab-1's Byron Lichtenberg, Michael Lampton (a Spacelab-1 backup payload specialist), and ESA astronaut Claude Nicollier. EOM-1 will use a short module packed with solar monitoring instruments and instruments to study the upper atmosphere. At first scheduled for December 1986, this flight was pushed forward into the fall of 1985.

EOM-2 may select new payload specialists or reuse the Spacelab-1 payload specialists, since the experiments are similar. It should fly in December 1987, under current flight manifest scheduling. EOM-3 will follow in November 1987 and EOM-4 in October 1988.

Spacelab 8, the International Microgravity Laboratory, will be devoted to materials processing, and the two payload specialists will probably be the backups for the prime payload specialists on Spacelab 3. It should fly in May 1987. A second such mission is scheduled for January 1989.

Two SHEAL (Shuttle High-Energy Astrophysics Laboratory) flights will each use two astrophysicists. The experiments include three x-ray astronomy investigations and one cosmic ray experiment, the last a reflight of a Spacelab-2 experiment for which payload specialists from that flight may be used.

Three missions called Sunlab 1, 2 and 3, exclusively dedicated to studying the sun, are due to be launched in mid-1986, mid-1987 and early 1989. Sunlab flights will require expert solar physicists aboard.

Spacelab 10 is a dedicated Life Science Laboratory, and two payload specialists, possibly the backups from Spacelab 4, may be used. It will be launched in June of 1988, with a follow-up a year later.

Although all these various scientific opportunities are the result of long years of advanced preparation, sometimes a payload specialist's mission may appear to materialize abruptly, apparently without the long years of intensive, carefully aimed effort. In October 1984 an oceanographer named Paul Scully-Power spent eight days on the shuttle just looking out the window — observing, describing, and photographing the world's oceans. He had been given a spaceflight ticket only three months earlier.

But the abruptness was not as real as it appeared. The Australian-born naturalized American oceanographer had spent years studying ocean photographs taken by astronauts and briefing and debriefing shuttle crewmembers on ocean dynamics. When the seat in space opened, on a mission already dedicated to Earth observation, NASA decided that Scully-Power's years of study made him one of the best-qualified oceanographic observers — on Earth or off it.

Originally, the backup payload specialists were not guaranteed a flight, but policy has shifted somewhat from the early days toward the inherent economy in using trained payload specialists, and optimism that opportunities for flying them will arise.

Scientific experiments will be features of virtually every shuttle flight from now on, but the use of so many experts so soon in the shuttle's regular operations bodes well for possibilities of using more scientists in the future.

According to Michael Sander, NASA is expecting to build up the launch schedule of Spacelab payloads to ten or eleven a year by the late 1980s. The various specialized scientific labs will be reflown frequently, offering scientists opportunities to repeat experiments, verify results, and adjust and upgrade their experiments from time to time.

After two decades of science from space, whole careers and entire scientific specialties may be built on the work of the payload specialist scientists who will be nothing less than the founders of future sciences. Like all the founders of past sciences, they will provide their colleagues with empirical proofs, with tangible products.

"I expect to become more of a spokesperson for materials processing in space," said Mary Helen Johnston. "If someone who's been in space and has done the work goes to conferences and publishes articles, the whole community of materials processing on Earth will take more interest in space. I think when we get this one system up and get good metals processed, people will realize it's not going to be five years away before we can do this — it's now. I think it will put space materials processing in the real world. Space was always seen by scientists outside the space program as something that strange people with way-out projects do. Now it's getting down to real practical matters. Industry is really the one that needs to drive it — that's when it's going to be really successful. If we can put a crystal on the desk and say, 'Look! This is the useful material! Look at the properties it has!' people are going to really get involved in it, in space, in space processing and space science."

5
The Military Manned
Spaceflight Engineer

"The military high ground is global," said former Major General John Kulpa, Jr. in a speech at the Johnson Space Center in 1982. "Space is now an essential part of national security: early warning, surveillance, weather survey, treaty monitoring, communications, and navigation. Space is important because there are things we can do better in space, and other things we can't do at all on Earth."

The Air Force space budget exceeded the civilian NASA budget in 1983 and 1984, an indication of just how big an interest the military has in space. The most expensive items were launch facilities at Vandenberg Air Force Base near Lompoc, California, which will enable the shuttle to be launched into polar orbit; a mission control center at Falcon Air Force Station in Colorado Springs, Colorado, from which future military missions will be directed; and very expensive satellites. For a while there was speculation that the Air Force might take over one of the four NASA Orbiters for military missions and form its own astronaut corps consisting of NASA-trained military astronauts. However, as it stands now, the Orbiters belong to NASA, the military is one of many customers NASA serves, and although the crews to Department of Defense (DOD) missions are always military officers within the astronaut corps, the astronauts themselves remain under the direction and control of the NASA flight operations directorate.

During the astronaut selections, the military prescreens its own officers for potential astronauts, both pilots and mission specialists, and forwards the names to NASA. These preselected officers then go through the usual interviews and medical exams as other candidates do. On the most recent NASA application, for instance, interested candidates from the service are required to send their applications, not to NASA, but to some person within their respective branch. Officers who are not selected in the prescreening process are discouraged from applying to NASA directly for the job. The one notable exception was black astronaut Fred Gregory, who was not one of the 133 pilots nominated by the Air Force because his experience was more in helicopters than in jet aircraft, and he had reached the age limit of 35. He submitted an application to NASA separately and offered to resign from the service. He told *Washington Post* reporter Thomas O'Toole: "I always think that if somebody puts that kind of limitation on you, that's the time to attack it. I never let things like that hold me up" (O'Toole 1978:4).

Gregory was allowed to stay in the service and was assigned to Spacelab 3 as the pilot.

Apart from the military people within the NASA astronaut corps, the military is also flying its own payload specialists on some DOD missions. Unlike the payload specialists from universities, these, called "manned spaceflight engineers" (MSE), are not doing scientific research. Unlike payload specialists from industry, such as McDonnell-Douglas' Charles Walker, MSEs are not uniquely expert in the operations of a piece of machinery. "MSEs do not individually design payloads," said the MSE program manager, a Navy commander who was involved with a navy space project for eight years before volunteering for the MSE job. "The systems are far more complex than the experiment designed by Mr. Walker and not subject to individual design. The MSE is part of the management team which oversees contractor efforts at system design and makes special inputs in the areas of shuttle interface requirements, flight design, and payload procedures development."

The MSE program was begun as an experiment in 1979 with 13 officers and was formally approved by Air Force Headquarters in December 1981. This group of 13 was described by General Kulpa as "development engineers who represent a focal point of knowledge on the shuttle, so the Air Force doesn't go down the wrong road." Their job then was to learn the shuttle inside and out. Kulpa described the group's activities as "getting together the first week of the month, unless it was a holiday, and going to Huntsville, Houston, or somewhere like that to become familiar with the shuttle. For instance, at Marshall Spaceflight Center in Huntsville, they put on spacesuits and went into the tank, just to get a feel for the mobility and maneuverability of it."

By the end of 1982, 14 more MSEs joined the original 13. Two of them are women and one is black. Only one had applied for the NASA astronaut selection in 1978 but had not been forwarded to NASA as a potential astronaut. The next selection of MSEs is planned for FY1986, with group selections to follow periodically as needed.

The MSE program manager oversees their training and assignments, gets briefings on their progress from time to time, and deals with the administrative problems and details of the dozen or so programs that fall in his domain. Although he would like to fly in space himself someday, he says "I probably never will." Under special arrangement with the Air Force, he agreed to provide exclusive information for this book, under the proviso of anonymity.

The Air Force did not put out a "cattle call" for MSEs as NASA typically does for its astronauts. Instead, the program is an elite "by invitation only" one. The Air Force personnel center screened their files for good candidates — officers with the right performance and technical qualifications — and then sent out letters of invitation to these individuals. Also, briefings on the MSE program were given at certain military bases and application packages distributed there. Applications were sent back, "normally with a commander's recommendation attached," after which a central selection board made its decision.

There were 285 applications received for both selections, out of which 27 finalists and 27 backups were picked. The backups, called "alternates," provide a pool of talent if any of the primaries should fail some medical requirements.

As with the astronaut corps, a bachelor's degree is mandatory, but advanced degrees are even more welcome. Only one MSE has only the bachelor's degree. Twenty-three have master's degrees, two have two master's degrees, and one has a Ph.D. Generally the graduate degrees are more varied than those in the astronaut corps. They include (the usual) engineering, aeronautical systems, astronautical, electrical, biomedical and mechanical engineering, general science, chemistry, physics, math and engineering physics, and (the unusual) radiation health physics, management business, systems management, engineering administration, and business administration. A flying background was not required, since the MSE is picked more for engineering and management skills than for ability to fly.

The finalists are made to go through the usual physical exams, which include blood and urine tests, resting and stress electrocardiogram, vestibular tests using rotating chairs and visual stimuli, visual acuity tests including color perception, hearing tests, echocardiogram, psychological profile battery, and psychiatric interview with an Air Force, not a NASA, psychiatrist. The psychological screening was, if anything, much more rigorous than that given at NASA. The brain/behavior tests looked not only for dysfunction but also for how well a person functioned. Short- and long-term memory tests were performed, spatial perception and memory examined, and pattern recognition examined in order to test rate of information assimilation, powers of abstract reasoning, and cognitive style. The MSE was then rated as unqualified, qualified with reservation, qualified, or well qualified by the psychologists, along with their reasons for the ratings.

Some other tests, not part of the NASA regimen, were also part of the physical: a 24-hour Holter monitor and a 24-hour sleep-deprived electroencephalogram. The rescue sphere was not used to test for claustrophobia, as it was for all astronauts and other payload specialists.

Also, unlike NASA, there were no final interviews. According to the program manager, "A central selection board, consisting of senior officers from the Systems Program Offices, the Manpower and Personnel Center, and the Medical Corps screened the applicants' service record and made a rank-ordered list of the top candidates. Program office experience of any kind (aircraft, satellite, subsystem) or satellite operations experience was preferred. An 'ideal' pattern would have a candidate with a technical bachelor's degree, four to six years of program office experience, and a master's degree in a technical or technical management field. Obviously, the candidate must be an outstanding performer in officer effectiveness reports. The real determinant or 'tie breaker' was proven performance in past assignments. There were more than sufficient numbers of applicants with the proper technical qualifications; we picked the 'go-getters.' "

In his personal opinion, the program manager rated performance on the job, aggressiveness ("self-starter"), self-confidence, perseverance, and technical competence as the most important MSE personal qualities. Certainly NASA would have roughly the same type of preference for its astronauts. Like NASA, the program manager rated lack of tact, brashness and egotism, stubbornness (let's do it my way), and not doing as one is told as distinct liabilities. "The MSEs chosen were considered to be team players," he said. "Personality traits not conducive to maintaining harmony in a close environment would be unacceptable." Graded performance on training exercises (something NASA does not do!) and the personal judgment of seniors counted heavily in determining whether an officer had the right stuff to be an MSE.

MSEs were, like the astronauts, generally physically active, and their hobbies included most of the usual astronaut activities: jogging, basketball, raquetball, skiing, scuba diving, weightlifting, flying, and auto mechanics. There were a model shipbuilder, two guitarists, a woodworker, and a person who played the recorder in the group, too.

Many had come from military families, and more than half had been through the Air Force Academy.

Age at selection varied from 25 to early thirties, with the average age about 31. All were officers ranging from lieutenant to lieutenant colonel with three to nine years' experience, the average being about nine years. A minimum number of about four years of service is necessary, because, as the program manager said: "We feel that four years is a minimum time for young officers to adequately demonstrate sustained superior performance; most MSEs (all but four at selection) exceed the minimum requirement."

However, this does not rule out sending an older officer of higher rank someday: 'There may come a time," the program manager said, "when a general officer may fly in space to assess program effectiveness, but it is premature to estimate when that might occur."

"The most impressive MSE to me, at least, is a 33-year-old major who was selected to that grade three years early," said the program manager — no mean feat in a peacetime military in which promotions are very slow and only spectacular performers are promoted even one year "below the zone" (early). "He holds a B.S. in electrical engineering, an M.S. in electro-optics, and has served as a missile crew commander and a shuttle project officer. He moves aggressively to solve problems, regardless of whether it's part of 'his job.' In addition, he keeps a low profile and works extremely well with other people and organizations."

MSEs will be, essentially, orbiting contract monitors. Their average day begins at 7 A.M. or earlier, involves going to meetings, phoning or meeting with contractors, developing procedures and tests for the payload, and meeting with NASA people working on payload-shuttle integration (making sure the payload works harmoniously with shuttle equipment and procedures). He or she leaves work at 4 P.M. usually later, and must travel out of town frequently, either to NASA, to contractor facilities, and so on, although home base is Air Force Space Division in Los Angeles for four to six years.

On the individual MSE project, the MSE may be a supervisor, a contract monitor, or just one of a group of people on a project, depending on the size and complexity of the project. Typically a project has one MSE assigned to it until two years before the shuttle launch, at which time a second MSE will be detailed to it. All MSEs work on one or more program, and several might meet for limited periods of time for special studies and design reviews. "There is no guarantee that an MSE working with a program will be the DOD payload specialist finally flying aboard the shuttle," said the program manager. "There may be some circumstances where it is preferable to designate a non-MSE as a payload specialist because of unique, detailed payload knowledge. But as a rule, MSEs do not individually design payloads."

The one-week-a-month shuttle familiarization will change over to a four-month course during which NASA and contractors will lecture over the area. After that, the training is on the job.

Training includes flight simulations at various contractor and NASA facilities. These include many different kinds of spacesuit and spacewalking activities. The MSEs practice working in a spacesuit in the water tank at Marshall Spaceflight Center, manipulation of the rocket-powered backpack called the "manned maneuvering unit," and the cherry-picker structure called the "Manipulator Foot Restraint" at Grumman. All these exercises will give the MSE an appreciation for what can and cannot be done outside the shuttle vehicle. In addition to this, MSEs typically visit the Rockwell Flight Systems Laboratory where shuttle cockpit procedures are worked out, and the aft flight deck facility, a simulator in which manipulation of satellites and other objects in the shuttle payload bay can be practiced. None of this is true specialized training to enable an MSE to fly the shuttle, to do the spacewalk, or to actually launch or retrieve a satellite using the shuttle's arm. "The purpose of these simulations has been to give the MSE an understanding of the shuttle capabilities and procedures. No MSE has participated in NASA joint integrated simulations, although MSEs selected as payload specialists [who will fly] will do so. The NASA-provided training will always include the minimum housekeeping and such Orbiter systems as may be required for payload activities." The latter is training virtually every payload specialist gets.

There are always certain conflicts that arise when a civilian agency like NASA must marry its procedures to a military one. NASA has always been a wide-open agency and has never had to give much thought to questions of security and limited publicity. Already there are changes to the facilities as a result of Air Force participation in it. For instance, access to Mission Control in Houston is now strictly controlled, both by additional security personnel and new stairways leading tourists in and out of the facility. In these types of matters, it is the military that can point out proper means of limiting access to previously minimally limited areas. People who work on military payloads, including astronauts (all military for DOD payloads), mission controllers, payload integration engineers, trainers, and so on must have the appropriate clearances. Most do already.

Larger questions also loom as potential problems. NASA and the Air Force have agreed that in cooperating on a technical level, NASA is responsible for shuttle operations, and the Air Force for its own payload, with "shared responsibility in the areas that overlap."

Beyond that there is a question of command. The MSE must

"always" serve Air Force interests in his/her dealings with NASA, according to the program manager. However, the question arises with regard to which commander the MSE answers to during a spaceflight — the NASA shuttle commander or the military bosses on the ground. "This is a delicate issue," said the program manager. "The crew commander must have control of his crew, including the DOD payload specialist. The payload specialist, however, is responsible to his military commander for the success of the payload-related activities. The exact details of how to mechanize/formalize the relationship and function are still being worked out."

As with any other piece of cargo, the Air Force will develop its own procedures and contingency plans, and the MSE duties specified in a timeline, just like most other payload specialists. However, in those contingency plans, there may be some decisions that the MSE must make in space, based on his/her on-the-spot assessment. Undoubtedly, the MSE, not the NASA commander, would have to answer for these decisions later.

Naturally, if the situation is life-threatening, the Air Force and NASA are unlikely to be in conflict. The dictum will probably be to save the shuttle and human lives.

The MSE program has been criticized, by some, as unnecessary. There is nothing about the payloads that the NASA astronaut cannot handle, critics say, and nothing really to be gained from sending an MSE on a spaceflight. The program manager defended the MSEs in this way: "The DOD payload specialist is necessary to provide in-depth program/payload knowledge; to act as security monitor/advisor from the standpoint of one who is accustomed to working in a secure environment; to be an advocate for the payload/program office in a situation where shuttle Orbiter problems or contingencies could impact payload operations; and to develop experience/expertise in manned spaceflight operations for future application to DOD/USAF requirements." In essence, the MSE who flies advises NASA on all security aspects of the flight, defends the integrity of the payload from NASA decisions that may affect the success of its mission, and must learn a great deal about manned spaceflight in order to design better military payloads in the future.

After a spaceflight, the current MSEs can expect to have "subsequent and more responsible assignments in the USAF Space Command and at the Space Division, where their experience can be best applied. Immediate postflight assignments will focus on 'lessons learned' to feed back to the program for future missions and to other programs for extraction of applicable portions for their own planning. As the involvement of the DOD in manned spaceflight activities increases, those with direct experience, like the MSEs, will be in a strong competitive position for the leadership roles." In essence, successful MSEs can expect promotion. But promotions for MSEs may not be all that fast and all that stupendous. "The near future may actually see some negative impact on developing careers," said the program manager. "But if things develop as I expect, that could well turn around, and involvement in the program would then provide a spur to future promotions and choice commands assignments." It is conceivable that MSEs might even suggest, propose, design, and even command future projects. It is likely that some MSEs will have active, not to say, brilliant careers: "They are a special breed. They are highly intelligent, aggressive, ambitious, hard-working overachievers."

The project manager did foresee a large MSE program in the future with "more opportunity for spaceflight." In ten years the program manager could foresee MSE payload specialists' doing a wide variety of jobs: operating complex payloads that are placed on an open platform in the shuttle payload bay called a "pallet," doing payload checkouts on deployable satellites, providing on-the-spot decisions when necessary, and doing routine maintenance and repair of satellites. Changes to military payloads in the 1990s will be predicated on the new shuttle vehicles that will be built to replace the ones being used now.

The people involved in the Air Force Space Division are, of course, motivated by a commitment to national security rather than scientific curiosity, as scientific payload specialists are, or by the aesthetics of the adventure, as observer payload specialists are. "Space exploration and exploitation is absolutely vital to national security and the continued survival of civilization as we know it," said the program manager. "We must build the capabilities to exploit the resources of the solar system before we exhaust what is available here on Earth. The role of the Air Force will be to defend the new frontier and the pioneers who go out to explore it, as well as to guard against any potentially hostile power who would use space to gain an insurmountable strategic, tactical, or economic advantage over the free world."

In 1985, the first MSE, Major Gary Payton accompanied the first shuttle-launched military payload into space aboard STS 51C. The payload, described by the Air Force as one which utilized an IUS (inertial upper stage) was of a secret nature. An IUS is a type of rocket that boosts large satellites from low Earth orbit to geosynchronous orbit. A rocket of this type malfunctioned in April 1983 when it failed to fire properly while attempting to place the very important TDRS (Tracking and Data Relay Satellite) into geosynchronous orbit. Fortunately, the propellant aboard the TDRS was sufficient to allow for its eventual placement in the proper orbit. The IUS has been carefully reevaluated.

Payton is described in *Who's Who in Aviation and Aerospace* as an MSE who has previously spent time as an instructor pilot and, later, as a spacecraft launch controller at the Kennedy Space Center. He was born in Rock Island, Illinois, received a bachelor of science degree from the Air Force Academy in astronautical engineering in 1971, and a master of science from Purdue University in astronautical and aeronautical engineering the following year (National Aeronautical Institute, 1983).

An insider in the space program described Payton's job as merely that of an observer.

Unfortunately, misinformed journalists have been speculating on the nature of the flight, and misinterpreting the MSE program in general. *Jane's Spaceflight Directory* claims that on STS-1, "USAF personnel parallelled and gradually began taking over many of the launch and mission control activities" (Turnill 1984:241), an assertion that is utterly and completely false. *Jane's* went on to claim that "NASA's traditional 'open' policy, with newsmen entitled to know everything that was going on, was gradually obscured" (Turnill 1984:241). In fact, the Department of Defense has not interfered with NASA press policy at all, except on DOD missions which carry classified payloads. For that matter, McDonnell-Douglas has been just as secretive about the drug that is being space-manufactured, the Germans will require proprietary secrecy about the "Navex" equipment aboard Spacelab D-1, and undoubtedly other corporations or nations may follow suit on future payloads. On the whole, however, NASA's press policy is as open as it ever was, and probably always will be.

The most ludicrous statement by *Jane's* was, "In addition to all this, there seems little doubt that as these activities advance, the military astronauts will soon be pioneering all sorts of exciting EVA work" (Turnill 1984:242). First of all, MSE's are not an astronaut corps in their own right but are merely a cadre of payload specialists most of whom will never fly in space. Secondly, spacewalks are currently performed to repair or retrieve satellites in very low Earth orbit. Most Air Force satellites are in orbits far too high for the shuttle to reach, and a majority of those are in geosynchronous orbit where the shuttle will never go and where spacewalks will probably not be done because of the high radiation levels there. Finally, manned spaceflight is something relatively new to the Air Force. Orbital reconnaissance, weather survey, communications and so on have always been exclusively the province of unmanned satellites. Even within the Air Force, there is little enthusiasm for manned spaceflight since unmanned systems have, thus far, carried out Air Force aims much better than a human could.

Jane's misses the whole point of the MSE program, which is to send observers into space for the same reasons corporations and scientific organizations do: to have someone "on the spot" in order to evaluate the progress of the mission, to troubleshoot, to alter an ill-conceived plan, and to get a feel for how space systems work so that they can be improved.

6

Foreign Space Friends

During the first two decades of spaceflight, Russians and Americans were about even in the number of astronauts they launched into space. However, because the space shuttle Orbiters can carry as many as eight people into orbit at one time and the Orbiters will be used many times in one year, the statistics will change dramatically. Of the next thousand people in space, the vast majority, perhaps 80 or 90 percent, will be Americans. James Oberg, the expert on the Soviet space program, said: "Until the Russians develop their own space shuttle, it is unlikely that they'll launch more than twelve people a year in the foreseeable future."

However, some people of other nationalities will be joining Americans in spaceflight aboard the shuttle, either because they purchased space for a large payload or because they are part of a joint scientific endeavor with us.

The first foreigner ever to fly into orbit and back on an American spacecraft was a West German, Ulf Merbold, who flew on Spacelab 1. He was born in the town of Griez during World War II. He graduated from high school in East Germany in 1960 and then defected to the West because he said: "I did not want to live under communism. People who spend their lives in the West do not understand how hard it is."

Defection at that time was simply a question of taking a subway train from one sector of Berlin to another, although getting into East Berlin from his town in East Germany was difficult, something he doesn't speak of even now.

He eventually settled in Stuttgart, receiving a bachelor's and doctorate in physics from Stuttgart University. He joined the Max-Planck Institute, first on scholarship as a student and later as a staff worker, involved in solid state physics. His specialty was and still is metals research.

Dark-eyed, dark-haired Merbold was one of four payload specialists selected by the European Space Agency (ESA) to fly on Spacelab flights. The ESA selection involved a widespread call for applications from scientists in participating countries. Merbold (a German), Claude Nicollier (Swiss), Wubbo Ockels (Dutch), and Franco Malerba (Italian) were the finalists. The Italian was dropped later because ESA felt it could afford only three payload specialists in training at the time.

Merbold spent no less than five years training for Spacelab 1, although he felt it was well worth the effort to play such a vital role in scientific experiments. And his role as "first" showed. He said he was always aware that Spacelab 1 "symbolized a new beginning for Europe." During the flight, Merbold did a wide range of experiments, including taking apart and repairing faulty furnaces, growing sunflowers, and chatting with West German Chancellor Kohl and President Reagan, tasks that underlined his scientific and diplomat purpose. He slept only three hours a night during the flight because he wanted to look out the window and take in the view.

Of the U.S. allies, Germany has shown enthusiastic interest in the space program. One entire Spacelab mission, called "D-1," has been purchased by Germany. In order to pick the payload specialists that would fly on D-1, the Germans asked the fifteen finalists for the ESA Spacelab-1 selection whether they were still interested in flying. Those who were went through the physicals.

The two men selected to fly on D-1 were both physicists: Ernst Messerschmid and Reinhardt Furrer. The German Space Agency also welcomed the participation of Wubbo Ockels, an ESA Dutch astronaut who acted as backup to Merbold.

The flight will have a record number of people on it: a commander to oversee the mission, one pilot and one mission specialist to respond to the necessary course changes (special maneuvers and pointing), two NASA astronaut mission specialists to help with the large number of experiments, and three payload specialists.

Much of the training will take place in Germany, except for the final simulations, which will take place in Houston.

Spacelab D-1, currently scheduled for late 1985, will carry scientific experiments in both the physical and life sciences. The Germans will be using the Spacelab module and the "SPAS," a platform that will carry additional experiments on it.

In the materials processing area, D-1 will repeat some of Spacelab-1's experiments and have some new ones of its own. These experiments include crystal growth, separation of nonmixable melts, soldering with microgravity, glass formation, and bubble transportation in liquids. In life sciences, there are several botany experiments testing geotropism. The gravity-sensing receptors of frogs will be tested. Human experimentation involves the "vestibular sled" to test how people sense acceleration in space. Biorack, which was on Spacelab 1, will run several cell experiments. The most unusual set of experiments involves time and space and is called "Navex." Part of the Navex experiment involves clock synchronization in which development and testing of a procedure of time distribution with the precision of less than ten nanoseconds is tested. The other Navex experiment relates to navigation, precisely determining distance in space.

West Germany is an open society and will publish the results of its experiments pretty much as NASA does. NASA will refly a few of the D-1 experiments on some of its other flights, too.

"We have very extensive cooperation with the Germans," said Cindy Cline of NASA Headquarters. "They're involved with many of our programs, including Galileo [a probe that will explore Jupiter], GRO [Gamma Ray Observatory], Rosat, and there might be another German Spacelab flight called D-4. We're looking at something called the Microgravity Lab, letting the Germans fly experiments on our facilities in exchange for their letting us fly some of our experiments on D-1. They have expressed an interest in cooperating with us on a space station too. Their cooperation would probably come in the form of swapping facilities, like on Spacelab."

Future projects that involve Germany include the launching of Eureca, an unmanned space platform that will be left in space for six months and then retrieved as a whole. Eureca consists of two SPAS pallet structures hooked together. The first processes to be parked on *Eureca* are microgravity furnaces. There is a current German-Italian study that is looking at parking a Spacelab-type module on a U.S. space station structure someday, too.

There will be many opportunities for flying payload specialists aboard these various cooperative flights, and it's possible that the ESA astronauts, Nicollier, Ockels, Merbold, Messerschmid, and Furrer may form a kind of European Astronaut Corps since training new participants is very costly and open selections are long and tedious.

The Canadians, like the Germans, were widely enthusiastic about getting their scientists into space and went all out in their own quest for the perfect spaceperson. The "Canadian Astronaut Program," as it was called, drew more than 4,000 applicants. In an extravagant selection process — which called for the selection board to travel to twenty different Canadian cities in order to hold interviews with potential finalists — six were selected, not only for their technical expertise but as national symbols. These six reflected Canada's geographic and cultural diversity. They included Robert Thirsk, a Montreal physician with a startling assortment of advanced degrees, including one in mechanical engineering; Marc Garneau, a naval commander with a Ph.D. in electrical engineering; Bjarni Tryggvason, an immigrant from Iceland; Roberta Bondar, a female physicist, and Kenneth Money and Steven MacLean, both of whom have an early and abiding interest in space. Of these, two will fly on the shuttle to test Canadian experiments involving the Canadian-made robotic arm (already installed on the shuttle) with an extended "machine vision" device to seek out and grasp objects. Marc Garneau flew on STS-41G in fall of 1984. Another payload specialist from the group will fly in 1985 and the third in 1986. It is possible that the others may get their chances in the coming years when Canada and the United States undertake more joint endeavors.

Italy has been invited to select a payload specialist on a cooperative mission between the United States and Italy. The flight is one in which the shuttle would deploy another satellite on a tether. Italy is developing the satellite itself, which is a group of sensors to measure conditions in the upper atmosphere. Its first flight is currently scheduled for 1987, and the Italians could select a group of individuals since the tethered satellite program may fly a number of times.

Spacelab J, which is due to fly in January 1988, may carry one or more Japanese payload specialists. Other projects with the Japanese, one involving a solar physics satellite, may provide an opportunity for yet another Japanese payload specialist in 1989.

Australia has always been very enthusiastic about space. They did not even seem to mind much when a hunk of Skylab fell on them, as other nations might have. Young and Crippen played "Waltzing Matilda" as they flew over Australia during STS-1, in homage to Australian enthusiasm. Two of NASA's key tracking stations are in Australia: the Orroral Valley and Yarragadee. Two Australian satellites are going to be launched in 1985, providing the Australians with two payload specialist opportunities. An astronomical laboratory called Starlab, scheduled for launch in 1990, will also provide Australia with opportunities for its space enthusiasts to fly.

Although the Australian Space Agency has not even put out a call for applicants, would-be spacefarers are constantly sending in their pleas to be considered. For the opportunities in 1985, Ron Goleby of the Australian Department of Science and Technology claimed that it would "be handy to have someone who would be able to provide us with first-hand information, bring back useful knowledge for Australian industry, and be able to relate his/her experiences with flair when he or she goes on promotional tours afterwards. It's a tall order." According to insiders, however, money is very tight in Australia at present, and the Australians may pass up the 1985 opportunities and aim for the Starlab opportunities.

The NASA policy toward flying foreign payload specialists has broadened a great deal over the last two years. Originally, a foreign payload specialist could fly if the foreign payload, say, a satellite, took up 51 percent or more of the shuttle payload bay. The language was revised recently to include any country that flies "a major payload" (that is, a payload of any size that goes into the payload bay but excluding customers buying only some middeck space or the very cheap "getaway special" space). Apart from that, many countries are undertaking cooperative scientific ventures with the United States, and this also qualified them to send expert payload specialists.

Under the new guidelines, payload specialists, from many countries could conceivably fly on the space shuttle. According to the most recent flight manifests, in 1985 alone, there are chances for payload specialists from ESA, Germany, Saudi Arabia, Mexico, France, and Australia. In future, the British, Luxembourgians, Italians, Indians, Indonesians, Colombians, Brazilians, Japanese, and Canadians all have opportunities to place one or more payload specialists on the shuttle. According to some reports, even the People's Republic of China may get a chance to fly one of its citizens. Some opportunities are provided by their major payloads, either government sponsored or industrially funded; in essence, some foreign payload specialists may come from foreign private industry, just like corporate astronauts in the United States, as well as from government-sponsored projects. For instance, five such slots are available to private companies in Great Britain before 1990.

Under such agreements, the foreign country or industry would have to foot the bill for the payload specialist's training, ranging from nothing to $100,000, sometimes more, depending on the payload and the training necessary to operate it in space. However, the selection of the payload specialist is strictly up to the government in question (except that the person or people selected must be able to pass all NASA medical and psychological tests and go through the training satisfactorily).

Although the wording of the current policy asks that the payload specialist in question have a "payload-related function," foreign governments need not necessarily pick a scientist or engineer payload specialist if it prefers not to. The NASA policy is currently so flexible that NASA might entertain any reasonable request, including flying a national leader or a great national artist. Other guidelines for that type of payload specialist would have to be used — perhaps those devised for the "citizen-observer payload specialist," which will allow U.S. journalists, writers, and artists to fly in space. "It would be exciting," said U.S. journalist Dave Dooling, "if Japan could fly one of its 'national treasure' artists someday."

7

New Firsts

Space is a stage upon which international competition takes place, national technology is verified, and national values are affirmed. For years it seemed as if the Soviets were sweeping some important firsts: the first man in space, the first woman in space, the firsts of many nationalities and races. The question of how substantive some of these firsts were is, of course, always open to question.

The Soviet Union is not above pulling publicity stunts for domestic and international consumption, and space is no exception. Although they flew the first woman, Valentina Tereshkova, very early in their program, they did not fly their second woman until nineteen years later. Tereshkova spent those nineteen years, not flying and reflying like her male colleagues, but giving public appearances, traveling around the world as an international celebrity, marrying in a huge flurry of publicity — to a bachelor cosmonaut, given away at the ceremony by at that time Soviet premier Nikita Khrushchev — giving birth to a daughter. Tereshkova shared little with her male colleagues in the way of qualifications, background and experience. She was not a flyer, had no military background, and had none of the "right stuff" of cosmonautics. She was, however, pretty, single, an active member of the communist party, a full-time factory worker, and a part-time parachutist. She was, moreover, Khrushchev's idea of Russian womanhood: wholesome, dedicated, patriotic, impeccably proletarian. In a conscious stroke of mythologizing, Khrushchev, himself a virtuoso publicity stuntman, understood that the label of "first woman in space" should hang on the slender neck of dreams, symbols, and national values.

The reality was quite different. Not only did the Soviet Union not fly another woman for nineteen years, but also Tereshkova was ostracized within the Soviet cosmonaut corps. One American, attached to the Apollo-Soyuz team, met Tereshkova at a party and said that she was shunned by her fellow cosmonauts. He chatted with her awhile and found that far from being the formidable statue of rectitude and propaganda, she was warm and friendly. He shyly asked her why she was so shunned by her male colleagues, and she said, quite plainly, "Because I am a woman and have invaded their playground." In fact, rumors that she had become hysterical while in space — never publicly affirmed — were widely circulated and used as an excuse for not flying another female for nearly two decades. Whether the accusation was true or was just a piece of malicious gossip cannot be known for certain.

When eight women, four blacks, a Japanese-American, and a Chicano were admitted into the American astronaut corps in 1978 and 1980, some critics complained bitterly that the United States was twenty years too late in allowing minorities and women into the space business. Members of previous selection boards gave a variety of reasons for this, the principal one being, as former astronaut Michael Collins stated: "No qualified women applied." The qualification issue centered around test pilot training, a predominantly white, Anglo-Saxon, male domain to this day. When NASA opened slots for a few scientist-astronauts in 1965 and 1967, there were still no minorities allowed in. (Although one black was allowed into the Air Force MOL program in 1967, this astronaut was subsequently killed in a plane accident, leaving the astronaut corps 100 percent white, Anglo-Saxon male.)

As for the women, according to some accounts *(Congressional Record,* June 1, 1983, p. 12), NASA did consider women for space travel as early as 1960. "Thirteen crack women pilots, including Jerrie Cobb, were put through the same tests as male astronauts and found to do well, but the project was dropped." Another opportunity opened in 1965 when scientists were first allowed into the program, with the condition that they learn to fly. Women were excluded from this selection as well.

In 1978 and 1980, the civil rights and womens' liberation movements had opened doors into universities, flight schools, and so on.

Talented minorities eagerly submitted their applications. And the NASA selection committees knew that racial and sexual equality had to play a part on the great stage of space. Along with other national virtues — courage, ingenuity, technological prowess, persistence, and hard work — justice had to play a prominent part.

NASA is quick to point out that these minorities were selected strictly on the basis of their qualifications, and in fact, the minority candidates were all brilliantly qualified: all but three have a Ph.D. or an M.D. Some have a long list of honors for academic, professional, or community achievements. They differ from the other astronauts in that more public attention has been focused on them individually than on other members of the space program. Not since the first moon landing has so much attention been placed on an astronaut as that placed on Sally Ride and Guion Bluford.

The press and perhaps the country sought to find stereotypes at first. Apart from the fact that the minority astronauts were all intelligent, there were few similarities in their early lives, their professions, and the paths that led them to the astronaut job.

The lion's share of the publicity seemed to fall on the women astronauts, especially the first woman of the group to fly, Sally Ride.

Sally Ride, unlike Valentina Tereshkova, defied every stereotype the media ever tried to foist on her, preflight or postflight. Although she is liberated, she is not a dogmatic feminist. Although she is a Ph.D. physicist and a thorough professional, she is not compulsively professional or scientific. Although she is a jogger, she is not a health fanatic, and she confesses to surviving on potato chips. Although she is a champion, she shuns the limelight. Although she is America's first spacewoman, she seems to reject the idea that she is a celebrity or a remarkable person. But like it or not, Sally Ride has had greatness thrust upon her.

For certain, some things are true of Ride. She does not wear her heart on her sleeve and is not forthcoming with personal information about her private thoughts, her private life. The best piece of journalism done about her was written by her childhood girlfriend, Susan Oakie. She enjoys being an enigma. Once, when a newsperson asked her if she would like to get married someday, she answered evasively, "Well, I won't do what Rhea Seddon did." (Rhea Seddon had married fellow astronaut Robert Gibson a couple years before.) The very next week, she did get married in a very private ceremony and in blue jeans: to fellow astronaut Steve Hawley.

She is, however, an absolute professional at her work. She was probably selected for her early flight for her mastery of using the shuttle arm, her quick intelligence, and her ability to function as part of a team.

Sally Ride was not a born astronaut. Even though she has had the honor of being America's first woman to fly into space, nothing in her background made that inevitable. Her father was not an aerospace engineer but a political science professor; her mother did not fly planes but stayed at home with her children, when they were young, and taught English afterward. Ride was not one of those youngsters who immerse themselves in science fiction and dream of a day when they would orbit the Earth. Her reading ran heavily into Nancy Drew mysteries as a girl and later into Shakespeare. Although she eventually majored in physics, she hadn't thought of a space career until she saw a NASA ad in her college newspaper calling for astronaut applicants.

In most respects, Sally Ride is no better than her other female colleagues. Any of them could have been "first" and honored the country. Of the 1978 class, all were Ph.D.s or M.D.s. For some — Shannon Lucid and Anna Fisher — selection was the culmination of long years of effort; for others, — Sally Ride, Rhea Seddon, Judy Resnik, Kathy Sullivan — it was a wonderful door that opened unexpectedly into a grand future.

The longest and most difficult road had been traveled by Shannon Lucid. She was incarcerated with her missionary parents in a Japanese POW camp in China when she was only six weeks old. They were part of a prisoner exchange when she was a year old and were returned to the United States. Lucid's great joy was flying, and her interest in space stemmed from her grade school days. "Even before Sputnik," she said, "I read exploration stories and science fiction, and space seemed to me a vast new place to explore." She saved up babysitting money as a teenager in order to take flying lessons.

Slightly older than the other female astronauts, Lucid had encountered some male prejudice along the way, in school and on the job. At the time she was picked for the astronaut job, she was married and had three children. Most women in this circumstance might find themselves in a difficult situation. The family would have to move. She would be putting in long hours on the job if she did get it. And her husband would have to give up his job and get another in Houston.

But the family stood behind her in the fulfillment of her own personal dream. When she was selected, her husband left his job to follow Shannon to Houston and found a job as a mining engineer for a petroleum company there. Lucid, a Ph.D. biochemist, has always worked, so her children have always been accustomed to having a mom with a job. "One day my daughter came home from a friend's house, and said in disbelief, 'You know, in some families, the mother doesn't even work!'" At home, everybody pitches in on the chores, and for fun, they go on picnics, go camping, or go flying. "My kids have flown since they were a week old," said Lucid.

Like Lucid, Anna Fisher had always loved space. According to her mother, Mrs. Riley Tingle, Anna always wanted to be an astronaut and talked with school counselors in order to find some way of getting a space job. At that time, NASA had no women astronauts, but Anna reasoned that she might, someday, find a place as a doctor on a space station. "She was always studying," said Mrs. Tingle. "Other kids would play and then work. She would do her homework first and then play." A self-motivated person, she seemed to show her capacity to learn quickly very early. "She put ice skates on for the first time," recalled Mrs. Tingle, "and just seemed to know how to skate!" Part of Anna's honeymoon was spent going through the NASA physical exams with her husband Bill Fisher, also an astronaut applicant. She was accepted in the first group in 1978, and Bill in the second, in 1980.

But great motivation was not the sole criterion. Sally Ride, Judy Resnik, and Rhea Seddon heard about the astronaut job more or less by accident. None had particularly geared their education to space, and some had not even thought of becoming astronauts at all until they read the announcement. And there were plenty of spectacular, motivated, educated women who had been turned down in 1978. There was Victoria Voge — one of the first women naval flight surgeons and the first admitted to the Navy's aerospace medicine residency program—who wanted to become an astronaut since she watched Captain Video on T.V. There was Dr. Mary Helen

Johnston, the first woman to graduate from Florida State University school of engineering. There were others too: Dr. Patricia Cowings, a psychologist, and Dr. Millie Wiley, a radiochemist, both of whom loved the space program. They were not turned down because they were not motivated, brilliant, exceptional performers. Nor will the astronaut selection board have the last word on whether these remarkable women will go into space. Cowings, Johnston, and Wiley (now Fulford) are familiar names in NASA literature, and Johnston and Fulford have been named as payload specialist scientists on Spacelab flights.

A few things had to change when women were admitted into the space program. The first was attitudes. Some of the older astronauts confessed to having a certain reservation about NASA's decision to allow women into the astronaut corps. "I used to think that being an astronaut was a man's job," said former astronaut Alan Bean. "But I was wrong. It's just as natural for a woman to be an astronaut as a man."

And it's possible that old attitudes die hard. A member of Congress pointed out that, of the 1978 mission specialists, four women were the last to be assigned to spaceflights.

There are still, inevitably, things women do that men will always have problems dealing with. One technician, responsible for packing the shuttle lockers, stowed 100 tampons for Sally Ride's short STS-7 flight because he had no idea how many tampons a woman using during menstruation and was afraid to ask.

Other changes occurred too, as a result of spacewomen. Personal hygiene kits, once stocked with male apparatus like aftershave lotion, boxer shorts, and razors, now include tampons, cosmetics, and female underwear. Spacesuits have been designed so that women can maneuver more easily. A "coed" toilet was also developed, and the all-male Skylab apparatus was scrapped. All shuttle equipment must be designed so that women as well as men can maneuver it.

The role of the spacewoman is still evolving. As a group, they fit no real female stereotypes and have to be treated as individuals. Seddon and Ride married after they came into the astronaut corps. Fisher and Seddon have both had babies since then, without impact to their job efficiency. Some, of course, prefer to remain single.

Sometimes chores, like shopping, have to be done at odd hours because of shift schedules, and sometimes family members have to pitch in. But these are small sacrifices for the chance to fly in space. Said Rhea Seddon: "I looked out the T-38 window the other day and asked myself, 'How many people get to have a job like this?'"

Like the women, the three blacks selected in 1978 were remarkable. Guion Bluford and Ron McNair were Ph.D.'s with a long list of honors. Fred Gregory showed stunning determination when he applied directly to NASA after being passed over by the Air Force for the astronaut job.

The first American black man to fly in space, Guion Bluford is a soft-spoken Air Force lieutenant colonel with a Ph.D. in aeronautical engineering. "Being the first black wasn't as important to me as just having the opportunity to fly," he said — almost exactly what Sally Ride had said during her first press conference.

His father, a mechanical engineer, encouraged Bluford's natural inclinations toward science and math and steered him in the direction of engineering. His interest in flying started very early, when he was eight years old, but it wasn't until he was in college that he actually learned to fly. Although his love was really aerospace engineering, he realized he could pass the physicals to become a pilot and believed, "I might be a better aerospace engineer if I were a pilot, so I went into the Air Force as a pilot." He flew fighter-bombers in Vietnam and was a T-38 instructor but flying was really a detour in his career: it was engineering he wanted to do. "I consider myself a technologist, and technology has always interested me."

He needed advanced degrees and was readily accepted into the engineering department at the Air Force Institute of Technology, where he received his master's and his Ph.D. He wrote his dissertation on a topic that interested the Air Force Flight Dynamics Laboratory, where his sympathetic commander left him alone to finish his dissertation.

Although he had no major mentors in school or in the Air Force, he did "have three or four people who have given me a push," he said.

Unlike Sally Ride, who claims never to have encountered prejudice, Bluford had. Bluford's children had been denied the right to go to the private base kindergarten. "I said, well, if they're on an

Air Force base, they ought to be able to select everybody," said Bluford. "So I went to a two-star general and he straightened that problem right out. There are times when you do that. There are other times when it might be wise to go around obstacles without a heavy struggle."

When he was first told of his flight assignment he almost didn't believe it. "I came in to work one Monday morning and was told after the astronaut meeting that I had a meeting with George Abbey [the Head of Flight Operations]. On my way over there I had thought I had done something wrong, but once I saw Dale and Dan [Dale Gardner and Dan Brandenstein, his crewmates on STS-8], I couldn't imagine the three of us had done something wrong. So I thought it was a project. We got into his office and George sat at the head of the table and said, 'I have to put together some crews and I want to know if you're all interested in flying.' We all were, of course. Then George, in his low key fashion, said, 'I'd like to put you on STS-8.' Our eyes opened wide, and it was a total shock to me — it caught me totally by surprise. Dale asked who the commander was going to be, and Dick Truly who was standing over to one side, asked, 'Can I be the commander?' and George said, 'Sure.'

"I remember riding down the elevator and thinking, 'Holy corn, I've been selected to fly! This must be a dream!' I was just elated."

The glare of media attention turned toward him immediately. "As I dove into the training aspect, a lot of attention was paid to my role as the first black and it was then that I decided how I was going to play that role. I wanted to have people look back at STS-8 and say, 'That was a good one, that was a fantastic flight and you did a really good job.' There was a lot of attention paid to me during the training phase or during the mission and I didn't want people to say, 'Well, he screwed up.' So I approached it from the standpoint of professionalism on the job, to do my professional best."

However, the media attention was hard for him. "I was uncomfortable with the public relations at first. I'm concerned about getting tired. I get less enthusiastic when I'm tired. I don't like to be on the road for more than two weeks; so as long as I can pace myself, it's o.k. There are both good and bad aspects about adulation. The good aspect is that it tickles your ego, everybody likes to hear that they've done a good job. But I bring myself back very quickly and tell myself that I'm just another person and one day this will go away quickly. And that day will come. Also, I can't live off of adulation. Adulation becomes tiresome because people want too much from you — they want your picture, your autograph, and they demand your attention. And they sort of ignore the fact that you need some privacy to yourself. And that becomes a bother sometimes, and I have to pace myself. For example, you go to a party and you're standing with your plate in your hand and you're eating, and they ask you for your autograph. Well, you're eating. And they do that throughout the whole affair. And you're still hungry and want to eat. Well, you think these people don't really care about you. So that's one of the bad aspects. People assume they can ask for you at any time they want to and forget that your time is your time — and then don't understand when you say, 'I don't want to give an autograph because I'm having dinner.'

"All this adulation will die down and you'll go back to work and be the same guy you were before you flew."

A self-acknowledged goal-oriented person, Bluford is already working on his next flight, Spacelab D-1. There were some very special moments he experienced in space. "The morning of the fourth day was very special. I woke up early and we were nose to Earth; we were on the dark side of the Earth. We watched an extraordinary sunrise. At that moment I thought how lucky I was to have an opportunity to see this. You do have some time to do some reflective thinking. I reflected on how lucky I am and looked out there and thought how beautiful the Earth was. I recognized I was there because of the efforts of a lot of people, as if I were the tip of the iceberg just above the Mission Control people, the trainers and so on. That's what the experience meant to me."

As the first black American in space, he views his role in history as part of a long chain of events. "I look upon my flight as an evolution. I can't ignore the fact that historically, along the way, people opened up opportunities. I recognize the efforts of those guys. Every year I talk to the Tuskegee Airmen and they're very proud of what I've done. But they were the ones who paved the road for me to do what I did, and I'm paving the roads for future black astronauts to do what they're going to do. I try to help people out who strike me as very talented. Help can be a word of encouragement or trying to open a door here or there if it's possible."

The same is true of the black astronauts as of the women: you have to take them as individuals. Fred Gregory is remarkable for his determination and his dedication to get into the space program. Guion Bluford is remarkable for the professionalism he brought to his task, his abiding and very genuine modesty, and the grace with which he wears the mantle of the first American black to fly into space. Ron McNair is remarkable for the diversity of his interests, his "Renaissance man" approach to the world.

A Ph.D. physicist, McNair is a master of the jazz saxophone and a black belt in karate, with a long list of academic and social awards. Over the years, he has been involved in a wide range of social programs, many of them aimed at kids. He was Cub Scout leader, a Boy Scout leader, and, for an organization of black engineers of which he is a member, a speaker to young black people, trying to excite their interest in the sciences.

Although he has many awards, he says, "I didn't need that kind of feedback. I was doing what I wanted to do the whole time. There was a lot of work, but you're learning, and the feedback is learning."

Success to him, "is being involved in something that brings satisfaction. If you're at the top of an organization and hate your work, then you're not a success. In the astronaut job, I'm always learning something new, new payloads, new experiments. It could be a sophisticated chemical process, a new camera — it could be anything."

Born of a family from a small town in South Carolina, McNair's parents never pushed him or his brother to achieve anything, although the kids all became college graduates. He had to play "catch-up" at MIT since his small-town educational background was not the best. But the years of struggle were "some of the best years of my life. It's an attitude thing." He did not aim his education at becoming an astronaut, but when the opportunity came by, he jumped at it.

"When the opportunity came by, I wasn't so sure it was real," he said. "I saw an ad which said, 'If you send in this coupon, we'll send you an application.' And the whole time I was tearing out the coupon, I kept thinking, 'This can't be real.' I sent it, and a few weeks later, a big stack of papers came back. When I saw the qualifications, I thought I had a good chance."

McNair's wife was always confident he'd get the job. "We were trying to buy a new car at the time," he said, "and we weren't rich by any means. An air conditioner costs $500, and we didn't need an air conditioner because we lived on the coast. But she said, 'We probably ought to get the air conditioner because we'll be moving to Houston next summer and it's hot there.' She was very matter-of-fact about it, and I had only just sent in the application!"

A physicist, McNair loves physically challenging sports, especially karate. "I enjoy mastering techniques, a proper sequence of movements. It's an art — making your body do certain things that required developing certain muscles, speed. It's physically challenging, mentally challenging. This became clear to me when I was doing research on physics and karate for an article. Physics and karate are very different, but they're two arts that require a lot of discipline; they both take a long time to learn, to reach a high level of skill — something you have to keep at in order to maintain a level of skill. The same is true of music. It requires discipline. There are a lot of things you can do to relax, feel better. But sitting down and doing physics takes discipline. Eighty percent of my karate students walked out because the long-term discipline of doing it, learning the movements, sweating, stretching, was too much. Difficult tasks, obstacles, road-blocks just make me work harder."

Ellison Onizuka, the first Asian-American astronaut, grew up in Hawaii. A third-generation Japanese, some traditions were kept up, including the celebration of both Japanese and Hawaiian holidays. "Values stressed were discipline, personal attributes, hard work," said Onizuka. "Education was very important."

Some of his family were interned during World War II. His grandparents ran a country store that was used as a local military headquarters.

Like some of the astronauts, Onizuka geared his education and professional life toward aviation and space. After receiving a B.S. and M.S. in aerospace engineering from the University of Colorado, he went into the Air Force, where he was involved in flight training at Edwards Air Force Base.

He has rarely encountered discrimination, but when he does, he usually turns the other cheek. "Most of society is very openminded about it, and it takes, I believe, small minds to foster discrimination. When you encounter it, getting upset doesn't help. You have to laugh at it or ignore it."

Although he is one of the "new firsts," he is bound to attract media attention. But his first crew assignment was, ironically, to STS-10, a top secret military flight that was temporarily delayed. "I don't think it was the fact that I was a minority astronaut that they picked me for that," he said.

He works with the Air Force on all aspects of the flight. "The Air Force and NASA have to work closely," he said. "There are differences between the Air Force and NASA, and those do have to be ironed out."

When Onizuka first came to work for NASA, he was on loan from the Air Force for seven years: two years of training and five years of operational work. "But because of the slips in the schedules," he said, "that time has been stretched out. We'll be extended here for a while and then go back to the military."

He views his role in space as helping in the "transition phase," transforming space from a frontier into a place people can work. "Aside from the fact that we'll have a few firsts, our generation of astronauts will transition the shuttle from occasional space shots to regular ones. We're making the scientific and technical community aware that space is viable. By doing this we're allowing the country to move onto bigger and better projects — building space stations, harnessing the sun's energy, mining the moon, attempting interplanetary travel. Those are all big projects, and you'd never be able to do that without a base to work from, and our astronaut class is creating that base. Ten years from now, I hope to be involved with NASA, working on a space station project or something like that."

He believes self-discipline, motivation, ability to work under stress, and humility are the personal traits necessary for a space career. "I think the most important attribute is integrity," he said. "You can't run a space program with people who are going to bluff their way through. Motivation is important because it's a self-motivating kind of job. I think humility is important too. We are no longer in the time where we can believe the world owes us something just because we are astronauts."

Franklin Chang, the first Hispanic astronaut, was selected in 1980. He is one of the most remarkable people in the space program. Born in Costa Rica, he knew he wanted to be a spaceman — and knew he had to come to the United States to do it.

"I was aware since I was seven years old that I wanted to go into space," he said. "I had always felt that the astronaut of the future was going to be a scientist rather than a pilot. I always associated space and rockets with complicated science and physics."

He got a great deal of moral support from his family. "My parents didn't think I was nuts. My mother was the kind of person who would never believe her kids were crazy. She would stand behind them no matter what they wanted to do. She is an inquisitive kind of person, and it is because of her that I maintained an interest in science. My father was more of an adventurer type. My father was a self-made engineer type, knowledgeable in maintaining heavy equipment for large projects like dams, bridges, and so on. He had to travel quite a bit for his work — Venezuela and so forth. I put the two things together: and in space, science and adventure are one. My father was the kind of person who was sure of himself; he knew there was nothing he could not do. He had so much self-confidence, and some of that spilled over to me."

He realized that if he wanted to become a scientist and an astronaut, he had to come to the United States. He moved in with distant cousins, poor people who welcomed him into their home. "When I came here I didn't speak English and didn't have any money. I lived with this family three or four months and did odd jobs. I went to public school to learn to speak English and to get a scholarship. Everything was a big adventure, and nobody had ever proven to me that I couldn't do it. Until I come up against a wall, I wouldn't think about not being able to do it. I was amazed sometimes that my classmates in high school were so insecure. They were so negative about things, and it's a state of mind. I felt that this country was great — I could see opportunity everywhere. And I couldn't believe why or how some of my classmates couldn't take advantage of it. They were all too afraid. Perhaps their families had sheltered them too much, maybe everything had been given to them. That's why I have so much respect for immigrants. This country is made of immigrants. They are highly motivated. By definition, an immigrant is a person that has to have a lot of motivation to move to a foreign country."

He had applied for the 1978 astronaut selection but didn't even get a response. "I always wondered if they even got the application."

Although there was some prejudice, Chang ignored it. "I had

instances of prejudice not very long ago, maybe four years ago. I had already had my Ph.D. and was working in the lab, and people in my neighborhood were prejudiced. It just comes and goes. When you see prejudice is rooted in basic ignorance, it is pointless to argue with that person. When people ask me if people in Costa Rica speak Mexican, I laugh and say they speak Spanish, because that's just basic ignorance. I haven't really experienced it from professors or supervisors. Most of those people were very educated, and educated people tend not to be prejudiced. But sometimes, when I encounter basic prejudice, I get mad and tell people they're bigots."

Selected as an astronaut because he can assimilate knowledge quickly, he is, in his own way, leading the way of transforming the role of the future astronaut: from a generalist who runs other experiments to a scientist who can perform his/her own research.

"When we have a space station, I can foresee where some of us who are scientists can play an important role in doing scientific research in space — not just do somebody else's experiment, but dreaming up and doing our own experiments as principal investigators. Things are always in the process of changing. We're in a situation now where we are pioneering new things, when people don't really know what we're going to do with, say, a space station. We talk about space station operations, and they say, 'Satellites!' But what are you going to do the rest of the time? They say, 'Well, materials science.' It would be very useful to have a scientist-astronaut right there doing the work instead of a p.i. [principal investigator] on the ground who really doesn't understand space. If the scientist is right there, he can change and modify the experiment. This is the concept behind the payload specialist, but you can have a mission specialist do it too."

Although Chang is flying in the face of the NASA dictum that insists that astronauts leave the scientific specialties behind and adopt the generalist approach to science, he is not worried. "I've gotten nothing but good feedback from NASA for my scientific research. I think that the management is interested in us keeping up our scientific proficiency. Training for mission specialist follows an exponential growth. It's fast at first because you're assimilating a great deal of information and learning the shuttle. It's just like learning a language. But as time passes, you pretty much understand everything there is to know about shuttle operations. You can learn more of it, to the utmost detail, or you can divide your time and pick up all the science you've left behind. That's been my case. I think if we're flying frequently, we'll basically be trained. Deploying a satellite is a standard operation; once you've learned it, deploying the next satellite won't be that much different. Other payloads are pretty automatic and don't require your complete attention. Right now there is an item called 'astronaut proficiency training.' It is geared toward astronauts who want to keep up their professional proficiency. They give us a limited amount of funds to carry out our research. I think this is important for mission specialists. We are scientists and we have so much to offer NASA."

Currently he is designing plasma rockets (advanced propulsion systems that will get us around the solar system swiftly) during evenings and on weekends. Once a month, he travels to MIT to check on the experiment's progress. "My work especially blends right into the space program," he said. "The problem with my plasma rocket is that we cannot test it on the ground. We must test it in space. We're trying to develop a small prototype. We'll put it on a pallet and use it like the SPAS [the SPAS was a platform that free floated close to the shuttle Orbiter during STS-7]. We will use the RMS [the "arm"], put it out into space, and let it fire to see how it behaves. Then we'll capture it, change it some more. This is the propulsion system that will get us to Mars in a few weeks."

Other space projects capture his imagination too. "Apart from my work at MIT, I took the original Boeing space station design and noticed that their habitat modules had no provisions for a laboratory workshop where you could do scientific work. They were laboring under the belief that experiments were already designed or controlled from the ground. They had not thought that the astronaut is also a scientist who can make changes. It's hard to design a laboratory on the ground, but once you have a multipurpose laboratory in orbit, you can do experiments and even do repairs to the station — a place to do basic welding, soldering, that kind of thing. I always believed you could run operations and do science at the same time, and they even nurture each other. So I designed a multipurpose laboratory, and it has been incorporated into the Boeing and Lockheed designs. Nobody ever asked me to do this — I just did it because I thought it was important for us to get involved in the space station because we're going to live there. For people who are going to be in the space station, you have to have an environment conducive to doing research. You could be working on materials science, for instance, and working it and working it until you get it right. Then you can automate it and put it in a free flyer. I think of a space station as a national laboratory. Maybe I am biased because I worked in national labs. There, you're not doing just science, you're doing applied science, trying to develop new things, new products."

And then there's his coffee maker. "I get interested in all kinds of things. Just last week I designed a zero-g coffee maker. I hate instant coffee; being from Costa Rica, it's like a sin to drink instant coffee. I felt like if I was going into space, I don't want to drink instant coffee. So I designed a coffee maker that could work in zero gravity."

Chang is excited by the future. "I am fascinated by life, by being alive, especially being alive now, being part of such an incredible adventure in a time when people go up into space. It's like being in the forefront of expansion. I'll go to Mars if I can. I'll have that engine ready!"

Chang is a rare blend of thinker and doer, perhaps of the same mentality of the great inventors. "I used to dream a lot about flying in space, going to other planets. I can't remember if I was driving. Maybe the spaceship was driving itself and I was just making sure it was going in the right direction. Now I think about very specific things, related to propulsion — I think about that. I try to see how these dreams can be made into reality. And I feel that now I have some very powerful tools to make these dreams into real things, to make them work. I'm really a very practical person. I went to school in engineering, but I didn't want to be an engineer constrained to blueprints. I wanted to be somewhere in between a physicist and an engineer, who thinks of things and then builds them."

8
Citizens in Space

A poet may write verse about a rocket firing in the blackness of space. An artist may capture the rosy glow of a fiery reentry. A novelist may write about how it feels to be alone in space among the stars and the planets. A philosopher may write the first treatise on human values in outer space. All this and much more may come to pass in the next decade as NASA opens up seats on the space shuttle to members of the humanities.

"Opening up space to the citizens was bound to happen," said Dr. Charles "Rick" Chappell, chief scientist for Spacelab 1, "because the shuttle is the step to the airliner which can take you and me, average people, into space. That means the spaceflight selection procedures will have to move away from the privileged few who have 'the right stuff.'"

Since the technical and piloting disciplines have thus far provided 100 percent of the world's astronauts, many people felt the "gee-whiz's" and "wow's" were not enough to convey to the American people the experience of spaceflight. Said astronaut Tom Mattingly during a press conference after the flight of STS-4: "I've given up trying to describe it. Maybe I don't have a command of the language. It's like describing a sunset—you have to be there." Many astronauts have expressed Mattingly's frustration in grasping for those fine, evocative words to describe the space experience — and failing: "Maybe someday we'll fly a poet up and conduct a press conference on the bridge of a spaceship and say, 'See, that's what we've been trying to describe.' And then the monkey will be on your back."

"The pile of letters kept getting higher and higher from the people who wanted to fly," said NASA Administrator James Beggs. There were letters from Bob Hope, John Denver, and Jacques Cousteau. There were neatly typed resumés and hand-scrawled notes. There were elegantly embossed envelopes from the rich and colorful, laboriously hand-decorated manila envelopes from the young. Thousands and thousands wanted to fly in space.

"Typically the way NASA reacts to unusual subjects like this in which we have established procedure or policy is to ask advice of the Advisory Council," said Beggs, who handed the difficult great historical question of which citizens should get to go.

The NASA Advisory Council meets regularly to advise the NASA administrator on a wide range of topics. The working group that was assigned the task of studying the citizen in space question consisted of Sylvia Fries, a historian; two public relations specialists, Florence Skelly and Julian Scherr; two engineers, Daniel Fink and Willis Hawkins; John Naugle, a former NASA chief scientist; and novelist James Michener. Astronaut Dick Truly was assigned to work with the group and offer advice on selection from the point of view of working space crews. The group sent out queries for opinions to approximately 200 experts and specialists and heard from many other people who volunteered unsolicited opinions and information. The task of the group was to study the feasibility of sending up a citizen, determine what kind of person would bring the greatest benefit to the American people, and set up some fair means by which this citizen could be chosen.

NASA made it clear from the beginning that there would be restrictions placed on the person who attained the glory of being the first invited guest of the space program. He or she would not be allowed on the first flight of a new shuttle vehicle and would not fly on a Spacelab or Department of Defense flight, and the flight would somehow benefit the American people. It would have to be a journey in which the individual would serve the public, not one in which he or she could take a joyride only or reap a huge personal profit.

At first, the names of the country's rich and prominent surfaced. "There's no reason why we couldn't fly the President of the United States," George Abbey, once told me, "if health and schedule permitted it." Former President Jimmy Carter, many members of Congress, noted explorers, world leaders, famous actors and actresses would provide the United States with a lengthy list of VIPs willing to fly. Prominence ensures that a person's career would not profit unduly from the experience, that a young person on his or her way up would not have a future "made" by the taxpayer-supported space program. "A person will be prohibited from making an extraordinary financial gain on it," said Fries. "That's to discourage commercialization of the program, so nobody gets extraordinarily rich on it. That would turn people off." A journalist could continue to draw regular pay but could not make a fortune on exclusives of his spaceflight. Similarly, a novelist, a musician, or an artist may be asked to plow extraordinary proceeds from the work resulting from the spaceflight back into NASA coffers.

But prominence was not the sole issue. There is much to argue against sending VIPs for their own sake. VIPs are people who are America's royalty, so to speak: they have been given society's highest rewards — fame, wealth, limitless opportunities — and many of them could eventually buy a ticket into space. The space program has, above all others, been the people's program and there is strong feeling both from within and outside NASA that the common person should be given the opportunity to fly in space. "I have maintained from day one that it should be a lottery," said historian Sylvia Fries. "The normal selection processes we have in this country are very institutionalized and tend to be dominated by cronies, and we aren't necessarily going to get the best person to do the job. As a historian, I have maintained that voyages of adventure have always been self-selected; they came forward and said, 'I want to go.' I think people within NASA are far less possessive of the space program than constituencies outside the space program, and I think there's far greater willingness to take the common Joe or the common Jane and give them a crack at spinning around and seeing what it's like."

But most agreed the time was not yet ripe to open up spaceflight seats to a lottery, in which a drum would roll, a number would be picked, and a person would go. The task was to bring some benefit to the American people, to find some way in which the millions of taxpayers who contribute to the space program every year could benefit.

"We were seeking people who can share experiences with other people, who can convey the feeling of flying in space," said Sylvia Fries. There are a wide range of people who fit into this category: journalists and media people, teachers, artists, musicians, moviemakers, poets, photographers, and so on. Each working group member favored his/her own type of communicator.

"Personally, I'm not all that excited about putting a great visual communicator or a great photographer up there. It would be hard to beat the photographs we have already," said Fries, "so I would favor someone in the verbal arts, someone who could go up and have that experience reflectively — maybe a poet, a novelist, or a journalist. The interesting story up in space is an ordinary person's reaction to the experience, what it's like to be out there in the dark."

"I wish we could fly a young Ansell Adams," said astronaut Truly. "Someone who can really capture what you see up there."

Daniel Fink, a former vice-president of General Electric and now a private engineering consultant, felt it might be "a writer who would obviously be able to express what it's like to be there better than a technician."

Former NASA chief scientist John Naugle recommended that someone from the reflective branches of the humanities — a historian, a futurist, a philosopher, or an educator — may be useful. "It might be a benefit to fly someone with a fresh viewpoint with entirely new concepts on the role of humankind in space, on what people might be doing 10, 20, or 100 years from now. This person might generate ideas that don't exist yet in the minds of people involved in the space program and might spring full-blown in the mind of a creative individual who'd say he'd like to fly."

James Michener thought it would be useful to have a filmmaker go up since cinema is now the most universal art form.

In January 1984, NASA sent to the *National Register* a request for a "communicator" who would briefly become a NASA employee for a period of about six months for the purpose of flying in space and telling about it. Interested individuals would apply for the slot, and intrinsic in the application would be a statement on why he or she should be accepted for this slot — a written account of the individual's motivation.

A committee made up of one's professional peers might sort through the applications, culling out those who are obviously not qualified or are in some way unsuitable for the job. Finalists will be made to go through the medical and psychological testing, and someone who showed "adaptability to the living situation and working relationships" would be considered." Finalists' way of handling emergencies may be tested in aerobatic flights in T-38s and in parabolic flights of KC-135s, or even in spacesuits at the bottom of the water immersion facility at the Johnson Space Center.

Undoubtedly, interviews with people involved in the selection and training will follow. The idea of selecting the prime and backup "observer payload specialists" as they will be termed, by lottery is unlikely. "The lottery system is too accidental," said James Beggs. "We're smart enough, I think, to work it out." Final selection will rest with Beggs and a committee of people involved in the flight.

By the end of this, NASA hopes to get someone who wants to go, could go, and could get along with people. As a NASA employee, the individual would be legally and ethically bound not to exploit the position for personal gain and to keep within the stated guidelines and procedures that all astronauts must go through. The person furthermore had to take directions from the command pilot, demonstrate the capability to "function as part of a small, highly knit team operating in a hazardous situation," have a proven or promising capability to communicate their experience to their compatriots, and be willing to undergo a full field investigation by the FBI.

James Beggs is technically the man to make the decision. A former president of General Dynamics, Beggs is a practical, fair, deliberate man with a flair for the tangible possibilities of space. The decision is not an easy one. "You have a potpourri of people who want to fly, and you must decide who can go," he said in 1983. "The first should probably be a journalist or someone who writes for a large circulation publication. A moviemaker making a documentary might be useful, I think. Whoever goes, our expectations should be modest. The first few years are unlikely to produce something terribly spectacular. The things that come to my mind when the astronauts come back is that the pictures don't do justice to the scenery, the spectacular view, the wonder of looking out, the blackness of space, the black background and the stars — all the soul stirring kinds of things. Astronauts are by their nature externalizers and basically trained to engineering and science. They don't have the inclination to look at things with an inner eye. There's a real experience there for someone who thinks in artistic terms to gain real meaning, ideally someone with a bit of the poet in him."

In September, 1984, it was President Reagan, not James Beggs, who decided that the first citizen in space should be a teacher. Although the choice bent some noses temporarily in the press corps, it turned out to be a logical, not to say, brilliant one. Education and space had been important aims of the Reagan Administration. Reagan emphasized excellence in education, especially technical education, in every way he could. He brought about a Renaissance in the space program almost single-handedly and against the warnings of some of his advisers when he committed the nation to the space station project and encouraged space development beyond that. Marrying education and space was a natural outgrowth of Reagan's themes.

The selection process follows the general procedure laid down in the *National Register*. Only full-time primary and secondary teachers are considered. Applications are to be sent to NASA Headquarters where a board of educators narrows the selection down to approximately 106: two from each state of the union, one each from Guam and Puerto Rico, two from the District of Columbia, and two from Department of Defense dependents' schools overseas and in the U.S. These 106 teachers are to undergo an orientation program at the Johnson Space Center, where they will be introduced to the space program and the education materials about it.

The 106 will ultimately be narrowed to 10 finalists who will then go through the physical and psychological screening. After that, a seven-member board of high-level NASA officials will meet, and then recommend the most suitable five. From that list, the NASA Administrator will choose one prime and one backup candidate, and flight training will begin. The final date of launch will depend on the availability of shuttle seats, either late in 1985 or sometime in 1986.

After the flight, the teacher will work at least a year as a NASA employee, making public appearances. In all likelihood, the individual will have to commit anything written or spoken about the experience to the public domain, and will be prohibited from making a profit on it.

In conjunction with sending a teacher into space, Reagan also encouraged the development of a program called the "Young Astronauts Council." Funded by private organizations, its aim is to push space and technology into the schools by developing curricula, improving the standards of math and science skills in the primary schools, using space information to heighten the technical awareness of young students, and sponsoring competitions in science and math for which the winners will be rewarded by going to see space launches.

In this way, Reagan and NASA hope to fulfill the first objective of the Citizens in Space Program: to disseminate information about space to the broadest possible cross-section of Americans.

But the working group felt that many people in the humanities would gladly pay any price to go on the great voyage of our time." "On the great voyages in the past, people were self-selected," said historian Sylvia Fries. "They said, 'I really want to go,' and they went."

The people most eager to go are the journalists. In the past, NASA was approached by an ancient but very eager Lowell Thomas, and by a somewhat younger Walter Cronkite, reporter emeritus now but still interested. Younger and less well-known journalists such as Dave Dooling, science editor from the *Huntsville Times* and Craig Covault of *Aviation Week* have tried to anticipate what NASA would like in a journalist and have gone through strenuous physicals and written about and experienced various aspects of spaceflight (KC-135, spacesuit training) to prove their worthiness. Journalists with long experience on the space beat have always wanted to go: Lynn Sherr (ABC), Morton Dean (INN), and Roy Neill (NBC) are familiar faces on television during spaceflights, as are Jerry Hannifin (*Time,*) Howard Benedict (UPI), Al Rossiter (AP), Thomas O'Toole (*Washington Post,*) John Noble Wilford (New *York Times*). Not so regular faces, but feature writers who have done good work in the area include Henry Cooper (*New Yorker,*) Rich Gore and Thomas Canby (*National Geographic,*) James Oberg of *Omni* and many others. They will all apply. The ranks of prominent journalists not specialized in aviation will swell the application numbers even more:

David Hartmann of *ABC's Good Morning America;* Diane Sawyer of the *CBS Sixty Minutes;* Tom Brokaw of the *NBC Evening News;* and Dan Rather of *CBS Evening News* have all shown interest in the slot.

For certain the first journalist in space will be a seminal figure in the history of American journalism. Implicit in the unprecedented opportunity of flying in space is the heavy obligation of making the story unique, fresh, and universal. The stories from the first space journalist will be the most scrutinized piece of journalism in American history, the most widely read or heard, the most vulnerable to attack. The journalist will be simultaneously the news and the newswriter. If the poor soul should use a hackneyed phrase, have a cliched thought, leave a modifier dangle, or utter the words "Gee whiz," the whole of the American press will be out there gunning for him or her.

Only two shields can protect the first space journalist from the thud, thud, thud of criticism. One is great prominence. The other is unique skill.

Space is a unique branch of reporting, and the question this committee will undoubtedly have to ask is: What kind of unique skill is needed to cover space first-hand? Those who would argue for a command of the language, a glitzy, evocative, alive style of writing or reporting would favor the true originals of journalism: Tom Wolfe, author of *The Right Stuff* or Charles Kuralt of *On The Road.*

Others argue that long experience in covering space stories is a unique skill.

What "the story" will wind up being is, in itself, a journey of exploration. We fully expect a description of the space experience itself: its awesome perspective, its special terrors, its disorienting rhythms, its immutable grace and beauty. Since astronauts are externalizers, people who are wedded to technology, the first space journalist may write of the first internal journey as seen by a person poised between science and the fine arts.

An appreciation for science as it is really done in space, a blow-by-blow description of physical and psychological adaptation to the profoundly alien environment, and accounts of crew interactions when the radios are shut down for the night all need to be told.

There will be restrictions placed on the story. They may come in the form of mandates forbidding reporting of certain events, like space adaptation syndrome in other crewmembers; they may come in the form of informal guidelines, gentlemen's agreements on what is or is not appropriate to discuss, somewhat like the *Life* magazine agreements with the original seven astronauts; it might come on the shifting boundary line of trust between journalist and interviewee, between freedom of the press and privacy.

Just as laboratory animals sometimes change their natural behavior in artificial, humanly manipulated environments, so also the astronauts may constrain themselves when being observed under the highly unusual circumstances of flying with a newsperson. Conversations may become polite, free of natural antagonisms, petty quarrels, swearing, and spontaneous expression, unnatural and boring. Their movements and behavior may change, knowing as they do that they are being watched. The great irony might be that the journalist, by his or her mere presence, may alter the story simply by being there.

Such problems may or may not be ironed out by the individual journalist. Some journalists are known for their combative, intrusive manner, their investigative method of reporting ("perfect candidates for a one-way trip out the airlock," said Dooling). The worry expressed by journalists unlikely to be first or even twentieth is that such an individual might ruin the chances for others. Others are milder, friendlier, sensitive to the ramifications of their reporting. "I would not describe how Sally Ride goes to the bathroom in space, but I think there's a funny story told in how I might," said Dooling.

Accuracy may not be jeopardized if one writes of altercations without naming names. One can be specific about biological processes without embarrassing individuals. In short, the writer can and probably will use some of the thousand special techniques journalists use in order not to violate an unspoken trust. "I would report on disruptive events which might endanger the mission or have long-range impacts on future crews," said Dooling, "but not the simple, unguarded phrases that might pass a crewmember's lips in the course of a hard day."

Most journalists would, however, bridle at undue restrictions being placed upon them. Although some journalists have been quoted as saying that they'd sell their souls for a chance to ride the shuttle, few would be willing to barter away their journalistic integrity. To force the journalist not to make a financial bonanza of his or her fortunate choice is one thing; to force a journalist not to make waves is quite another. "Some upper management types are hypersensitive," said Dooling. Most journalists would fight any effort to obstruct a legitimate story. More crucial to the story is the difference between individual journalists — how they handle trusts, how they write, what bias they have — which will all come into play when writing the story of the century. NASA's choice of the individual journalist will, in part, presuppose the type of story that will be written. Whether the prose is tightly precise and highly technical or openly evocative and poetic, whether it represents the news-writing norm or the highly unique divergent stylist, whether it is written by an externalizer or an internalizer will undoubtedly set some kind of standard for space journalism with which future journalists will have to compete.

It is, however, unlikely that the first individual journalist will set some rigid personal and professional criteria for future journalists. The individual may be uniquely prominent or uniquely skilled, and NASA may decide that while it likes homogeneity in its astronauts, it dislikes sameness in its journalists. NASA could, and perhaps already does, believe that the American public is best served by sending up a wide diversity of "communicators," with widely differing viewpoints and skills.

Young writers such as Dooling and Covault have dedicated their journalistic lives to space and may be more representative of the typical space journalist in the years to come — the 1990s and beyond. Covault has written by his own estimation "over three hundred" articles about the space shuttle. His understanding of the spacecraft's operations is so deep that he frequently stumps astronauts with his very technical questions at press conferences. "He is accurate!" claimed one astronaut. "He's certainly one of the top twenty I would consider to fly in space."

Covault came from a small town in Ohio not far from Wright-Patterson Air Force Base. He always loved aviation and space. He even tried to pass the physical to become an Air Force pilot, but his eyesight was not good enough.

While he was in journalism school, he specialized, as much as he could, in science and aviation writing. He knew he wanted to write for *Aviation Week* since he was 17, in order to write the kinds of stories he wanted to write, and interviewed with them in order to find out what qualifications he needed. A *Reader's Digest* grant allowed him to get close to the space program. During his junior year, he got to go to Cape Canaveral and work in spacecraft simulators with astronauts. The following year, he went to the Johnson Space Center and did the same thing. "At Cape Canaveral, I watched Gene Cernan fly lunar landing sims. Also, one highlight was spending a day and a half in the manned spacecraft operations building in the spacecraft checkout area watching lunar modules and service modules getting ready to go to the moon. The next year, at Johnson Space Center, I worked in simulators by myself and ran checklists, and then interviewed astronauts. The first interview with an astronaut I ever did was with Buzz Aldrin."

While he was a junior in college, he learned to fly an airplane.

Apart from journalism school, he needed at least a year as a working reporter on a newspaper. After his year was up, he went back to *Aviation Week*, where he wrote a wide range of stories on various aspects of the American and Soviet space programs and on high performance test aircraft. Going into space was always a goal.

He feels that writing those hundreds of articles on the shuttle has educated him considerably. "If you want to do something, you ought to prepare yourself for it," said Covault. "You shouldn't depend on being an anomaly. The old American way is to set a goal, do the hard work and do it. As a journalist, you have to understand how these things are done."

Dave Dooling writes for a somewhat smaller audience but has a similar reputation for accuracy and has branched out into magazines and books to extend his writing range. "Apart from the physical and psychological qualifications," Dooling said, "I think a space journalist should have certain professional ones too. I would look at accuracy, breadth, and depth of coverage; a certain number of years in the area; stature within the writing community. The journalist must be good at space and be willing to learn. Like payload specialists, they bring specialized skills to the job."

Dooling's career is quite different from Covault's but typical, in a way, of a certain breed of journalists who are not educated to the craft but who train themselves, have a flair for the work.

An engineering school dropout, Dooling was bitten by the journalistic bug when he was working as a night watchman. "I read a magazine article which compared the lunar lander to von Braun's old concept. But I had read lots of magazines about it and had witnessed the history of lunar lander development, and I wanted to write about it." The result was a verbose 25-page article that finally ran in the British magazine *Spaceflight*. In the process of preparing the article, he learned to research, learned to organize. He worked briefly at various newspapers, taking criticism, learning about deadlines, working as copyboy, reporter, editor, and correspondent. Although he was, at various times, writing news, crime, agriculture, and cutesy features like the one about the crow that smoked and ladies in the swamps doing ecological research, his love was science and space. In 1977, he became science editor of the *Huntsville Times*, close to the Marshall Space Flight Center, where he devotes almost his entire career to space writing. He was also editor of *Space World* magazine.

Although Covault and Dooling are as different as two people can be, they share certain qualities that will be typical of space journalists in the future: accuracy, depth, breadth, love of space, dedication, discipline, and good health.

Whatever the qualifications, one great service the journalists and other communicators will provide is making it easier for common people to get into space. They will do so by dispelling the myth that nontechnical people might not be fit to handle the disorienting aspects of spaceflight. For nearly two decades the argument against flying communicators rested on the question of sanity. The measure of predicting future behavior was to look at past behavior, and it seemed that pilots or people who have been in life-threatening situations had truly proven their sanity by exercising cool judgment in the face of death. But very few communicators are pilots. The anxiety that a nonpilot, nontechnical person "could go crazy" and "endanger the crew" still surfaces. This last bastion of prejudice — perhaps the last excuse—that keeps the common citizen from going into space is falling and will crumble completely once the first communicator leads the way (unless the "communicators" are all required to be pilots too). "I'm sure Magellan didn't take any kooks with him, but that raises an interesting question," said author James Michener, one of the members of the NASA Advisory Council. "Was the sailor on the third watch on Vasco da Gama's trip essentially more stable than the poet he took along? Is the guy who's going up in space going to be more stable than me? I doubt it seriously."

For journalists, the ultimate result of the space trip will be a unique piece of work. Whether it will be truly great is another matter.

"I think we're playing the averages with a space writer," said Michener. "I can think of 20 people we could nominate, and a few would be very good and 14 or 15 might be very ordinary. Nonetheless, we're not going to leave this in the hands of the technologists.

"There will be a division between the earthbound and the space-bound someday, but I think the effects will be reported, and writers will be conversant with them. Artists and writers will have a profound effect on the future."

It is possible that the role of the space journalist will evolve, mature, and possibly transcend journalism as we know it now. The reporter may become the interpreter, the interpreter the philosopher, the philosopher the prophet. Future writers, said Michener, will "lock into the human consciousness that we do live in this little envelope, and beyond this there's this new universe. We thought in a universal sense after Copernicus, but then only 5 percent of the people thought or knew about it. But it'll be different with space. Because so many more people were exposed to the space program, we might have our consciousness expanded at a very fast rate."

For better or for worse, American journalists have been on the frontlines of wars, economic and social turmoil, medical and technological breakthroughs, and simple everyday life. Finally, a journalist will step across the threshold of the known world into space and will attempt to write for us a new *Odyssey*, the ever-new, never-changing urge to reach beyond the known world. The first journalist will be standing ready, as journalists have stood on past battlefields, past thresholds: courageous, truthful, full of wonder at the human condition.

Artists may be next. In discussing the right artists to be picked for the slot, Robert McCall has been mentioned. His murals grace the Air and Space Museum, the Johnson Space Center, the Dryden Flight Research Facility. His space paintings for books, magazines, postage stamps, and private collections have been regarded as the best of the genre.

"I think art is an adjunct to photography," he said in an interview in the Johnson Space Center newspaper, *The Roundup*, "another way of documenting and recording historical events. An artist can inject an emotion into his work. For example, in the mural at the Johnson Space Center, the sky is bright and inviting, not black."

He feels the decision to fly artists in the future is a good one. "The astronaut personality is fairly stereotyped, although as you get acquainted with a lot of astronauts you recognize they are very diverse. But essentially, the creative temperament has really not been the temperament which makes astronauts . . . but when the first group of creative people go up who are able to look for different things than the engineer pilot type of person, we will be hearing different things about space. Maybe some things which are described in a fresher, new way that the masses can relate to." McCall believes that when space stations become large enough, maybe up to 30 residents, artists may be allowed a brief time to paint the environment. "For my own part, I would like to set my pallet up just above the rings of Saturn and out about ten thousand miles or a little farther from the surface, adjacent to one of the moons. I think that would be a fantastic place to make a painting."

Ex-astronaut Alan Bean, fourth man on the moon and Skylab 3 commander, turned artist after leaving the astronaut office. "Art is important in comprehending space," said Bean, "because the photographs can't duplicate the colors, the textures of being there."

Bean, who is now pursuing a very successful career as an artist, started to paint in his spare time while working first as a pilot and later as an astronaut. He is still haunted by his few days on the moon. "I want to be the impressionistic painter of the moon," he said.

"I go over the color reports and if they say this is a tan gabro or light green gabro, that's what I paint. My goal is to paint the moon as I saw it, something that is true to me. The artist loves it and paints it."

Placing a photo of a moonscape next to a Bean painting of it, the moon looks softer, less angular, less hostile.

"It's very beautiful. Every astronaut up there has said how beautiful it was yet the pictures of it make it look cold, hostile. I'm going to paint the moon as beautiful as I remember it or as other people have reported it to be. Since coming back from the moon, I look at pictures a lot and even those pictures have begun to look different to me, more beautiful. Perhaps I now see possibilities for beautiful paintings."

Musician John Denver and filmmaker George Lucas are eager to go into space, as are many other people from the humanities. Undoubtedly, in the next twenty years of spaceflight, these creative people will have their chance to go into space.

It is possible that the most exciting possibilities for the arts lie not in known art forms but in new ones. Molten metal and liquid glass will probably displace paint and canvas as space art materials and may catapult glass and metal sculpture into the primary art forms of the 21st century.

Opportunities for fine artists will pave the way for space tourism, which will undoubtedly be one of the leading industries, if not the leading space industry, in the 21st century. Although Pan Am and TWA once took reservations for the first tourist flight to the moon and several companies are trying to do the same for the shuttle, there really is no officially sanctioned means of having those reservations materialize into real tickets. However, by the 1990s, a hotel module could fly into space in the payload bay of the space shuttle or passengers could be carried to an orbiting hotel complex.

"Ultimately spaceflight, a sojourn to a Holiday Inn in space, will be part of people's vacation plans," said John Naugle.

A double-decker passenger module was designed by Rockwell that could carry 74 passengers into space. It included emergency exits, its own environmental control, air, water, life support, and comfortable seating but no toilet. "Fortunately the ride would be short. Passengers would board a few hours before launch and disembark a few hours after reaching orbit. At that time, early concepts called for using the shuttle as an airliner which could transport people from New York to Tokyo in 45 minutes," said William O'Donnell of NASA Headquarters.

The same module could be modified and used to transport people to a space hotel, but "they wouldn't be able to see the takeoff and landing because the shuttle payload bay doors must be closed then," said O'Donnell.

Another concept called for linking together two Spacelab-like modules as a passenger carrier, without the scientific equipment, of course. It was suggested that half of the seats would be for sale at a high price while the other half would be raffled off in a lottery fashion.

In order to have a passenger module that could serve as a hotel, as well as a transportation vehicle, it would have to include some of the amenities and would probably carry fewer passengers, maybe only 40 or 50. "The cost of a seat would be prohibitive," said O'Donnell, "something like 2 million dollars a seat."

Undoubtedly serious tourism entrepreneurs would find some way to bring that prohibitive cost down eventually to something within range of more moderate incomes.

"It's crossed a number of people's minds," said James Beggs, "that there may come a time in the 1990s when there is a desire and a market for tourism. The question is, should the experience be permitted only to people who have the money to fly? As a paying proposition, we'd have to look at it. If you design a module that has windows, or something like a glass bottom boat, the cost would not be insignificant. It would have to fly over a period of many years before one could get a return on the investment. If I were a businessman, I would probably want to wait until the first nonastronauts flew to get a feel for what they really enjoyed, what they thought was important, and then design a capsule or module with their remarks in mind. I believe that in the 1988 time period, a ticket may cost less than a million dollars. I think an orbiting hotel might come later. If we go ahead with a space station, we could probably build a hotel in space and then, on a space-available basis, we could take tourists up. I can see that happening in the 1990s."

"With the Air and Space Museum in Washington, D.C., being the most patronized museum in the world and with the splended book by Tom Wolfe [The Right Stuff]," said James Michener, "I think we've had quite a lot about space filter down into the mass population. I think one of the most profound things of my lifetime were those first pictures showing that the Earth really is a globe, and it's really floating in space and has a fragile atmosphere. I don't think people have gotten over that yet. It was one fantastic intellectual breakthrough. And when I wrote *Space*, I was impressed with the tremendous complexity of the effort.

"I have been saying for the last two years that we could put up people who simply wanted to go today. Unless people like me are keeping attention on that, it will become the plaything of the scientists, and it would be a dreadful loss if that happened."

9
The Corporate Astronaut

American businesses were, at least in the late 1970s, shy about investing huge capital in space business — and with good reason. A company would have to build and design production equipment, test the equipment and the product both on Earth and in space, and sell to a known market. For such a venture, launch costs alone would overwhelm the potential investors, who could, after all, make quicker profits in other places.

In 1979, NASA developed a policy that would encourage these infant industries. It signed joint endeavor agreements with a few corporations interested in space manufacturing. NASA would fly experiments and space processing equipment and provide technical support free to any company that looked as though it would someday become a regular paying customer of the shuttle. The result was that companies could think more seriously of developing space as a business frontier without worrying, at first, about the huge launch costs involved in space payloads.

The end result may be the most far-reaching one in our economic history. It may create the industries that may eventually be the basis of new wealth and new fortunes for an eager world.

Pharmaceutical production, metallurgy, and crystal growth were acknowledged, at that time, to be potentially profitable space industries. It was precisely the businesses in these industries that signed the first joint-endeavor agreements.

Of these, pharmaceuticals production will be the earliest and most profitable space manufacturing industry. Although drug manufacture had been identified in government studies as a prime space industry as early as 1969, it took a psychological breakthrough in the business community to bring this speculation to reality.

The leap of faith was first taken by McDonnell-Douglas, a large aircraft and aerospace firm used to dealing with the future.

Studies had shown that of all pharmaceutical processes, electrophoresis had the most to gain from the space environment. In continuous-flow electrophoresis, biological material is injected into a fluid being shot through a long tube. An electrical field is then applied, pulling the materials into separate streams, after which they are collected. In gravity, the biological material is pulled down to the bottom of the tube by gravity, and the electric field induces heat that, on Earth, leads to convection (hotter material rises up), mixing the separated streams. Because space is a weightless environment and convection problems are nonexistent, this process was targeted in several studies in the 1970's as the one most likely to benefit from space.

Project manager James Rose, an engineer who had been involved with NASA since 1957 (back when NASA was NACA), left his job as director of engineering at NASA headquarters to head up what was later called the EOS Project. The term EOS is a double entendre. It is an acronym for "Electrophoresis Operations in Space," but it is also the Greek word for "dawn" — what McDonnell-Douglas advertises as the "dawn of a new era in space." And Rose was just the person to head up this kind of project. Grounded in spaceflight engineering, he was an adept manager who had directed several other McDonnell-Douglas projects and was known for his logical, very methodical approach to new ventures.

The approach McDonnell-Douglas took was, at first, very conservative. Rose focused attention on electrophoresis because of all drug manufacturing processes, it stood the most to gain from the space environment, especially in the amounts and types of pharmaceuticals that it could yield. Also, McDonnell-Douglas had the only active life science group (made up of physicians, microbiologists, and so on) in the aerospace industry because of their involvement first in Mercury and later in Viking. "We put the biologists and engineers together so they could talk to each other, and they have ever since," said Rose.

The engineers generated mathematical models of the fluid dynamics and mechanical aspects of the problem and later built the hardware. In 1977, McDonnell-Douglas sought a partner in the venture. McDonnell-Douglas wanted a drug company to take over the FDA approval and the marketing of the drug or drugs yielded by such a space project... and to share the risks.

At that time, the shuttle had not yet even flown, and most companies were looking to invest in shorter term and more secure endeavors. When McDonnell-Douglas approached some drug companies, the response was, "That's not going to happen until 2040." It took another leap of faith, and it was Johnson & Johnson that took it. The agreement between McDonnell-Douglas and Johnson & Johnson was signed in 1978. Under the terms of the agreement, McDonnell-Douglas agreed to produce a few products exclusively for Johnson & Johnson but remained free to enter into agreements with other companies to produce other substances (as it did with Washington University, producing beta cells as a possible one-shot cure for diabetes). NASA is also allowed to use the electrophoresis equipment to experiment on its own research materials. As soon as the EOS project ceases to be experimental and goes into the commercial production mode (that is, when it generates drugs for sale and not for test purposes), the corporations will pay NASA for the cost of the launches.

In what can only be termed "fate," a young engineer's resumé crossed the desk of the fluid subsystems department manager at McDonnell-Douglas at the same time James Rose's request for a CFES (Continuous-Flow Electrophoresis System) engineer did. Rose required someone with expertise in fluid dynamics and astronautical engineering, with a test engineering background, and with a strong interest in space. This individual would do both basic research on and construction of an operational device. The young engineer selected was a test and design engineer for the U.S. Navy who had degrees in aeronautical and astronautical engineering from Purdue. His name was Charles Walker.

Walker was what might be called a "space nut." When Sputnik was launched in October 1957, Walker was in the fourth grade. There was a big bulletin board in the classroom, and his teacher asked everybody in the class to bring in pictures and articles about rockets and satellites. "I can still remember those pictures and articles on the bulletin board," he said. "I demonstrated such a strong interest that she gave me all those articles and clipping after the unit was over, and I kept them in scrapbooks. I still have them." He also read the seminal works of Wernher von Braun, who made the future so tangible and far-flung ideas so possible to young people. Besides that, he gravitated to the science fiction authors who "clanked": Isaac Asimov, James Blish, Arthur C. Clarke, all well grounded in science and who, like von Braun, rendered future technologies believable and tangible.

In 1957, he and three other boys in Bedford, Indiana, built solid rockets using aluminum tubes from T.V. antennas, kitchen matches, and paper clips. Occasionally the rockets would hit houses, and the boys got some bad press, but they kept building bigger and better amateur rockets, using gunpowder-based and eventually bound complex prochlorate propellants. "The cheap way to bind up a propellant was to use caramel — that's right, melted caramel as an oxidizer," Walker recalled.

The work was overseen by Walker's father, a movie projectionist by trade and an engineer by avocation. He had learned mechanical, electrical work and carpentry during the depression and World War II and had a full machinist and woodworking shop in the basement of his house. Walker's mother, an acknowledged tomboy, pretended not to approve of the rockets' hitting people's houses, but the father assured the boys: "Your mom doesn't like us doing this, but she'd love to be out here too."

When Walker graduated from college in 1971, the job market looked grim for a space job. And there were temptations from other quarters. Environmental activities were very big in the early 1970s, and Walker had been involved in those from the late 1960s, when he worked as a fire fighter during summer vacation from Purdue. He also worked as an engineering technician for the forest service and, in 1970, was offered a supervisor's position as an environmental resource manager after graduation. "The sales pitch was: 'There's grass roots activity here, and there's going to be more legislation and more jobs in this area. Someone with a technical bent has a lot of potential here.' I have a certain amount of dedication to improve the envirorment for future generations. But I said, 'No, the wheel will turn, and jobs will come around in space.' I was willing to tread water until then."

After working at Bendix Corporation and for the U.S. Navy, he was hired by McDonnell-Douglas. He did apply to NASA to become an astronaut in 1978 but was turned down.

Walker worked on another McDonnell-Douglas project briefly, until James Rose pulled him into the CFES project. For three years, a tight core of four individuals worked 60 and 80 hours a week building the CFES. They developed a generic process of electrophoresis—that is, one mechanism that could process a wide variety of products. On the first three flights of the CFES aboard STS-4, -6 and -7, protein materials in solution were processed. On STS-8, live cells were processed.

Although the first CFES experiments were performed on a machine that was semiautomatic, the scale of the project was small and it was certainly not enough to generate the quantities necessary for animal and human tests required by the FDA. But flawless equipment performance on those spaceflights did accelerate the project considerably. Walker redesigned the CFES in order to get it to function for more extended periods of time in orbit, generating enough product to begin animal studies with. However, when Walker and the CFES group redesigned the CFES to scale up to a machine capable of producing larger quantities, it was difficult to automate it. And it would have taken an astronaut several hundred hours of training to gain enough expertise to work the CFES in that mode. "The process requires visual observation in order to fully comprehend the quality of the separation. There's a lot of art to that. There are some visual qualities, some integrations of other parameters like temperature and pressures and so on. We don't yet have an algorithm for it — in order to train someone quickly on it or to get it into an automated system."

When NASA was confronted with the prospect of tying up several hundred hours of an astronaut's valuable time training on one device that might fly only a few times in the next several years, NASA agreed that it was more expeditious to send Walker himself. That was, after all, the definition of a payload specialist: someone whose unique capabilities require his presence onboard the shuttle to oversee a complex experiment.

NASA agreed to fly Walker, and McDonnell-Douglas agreed to pay for his training.

Before the flight of STS-12 (now called STS 41-D), Walker not only worked on the design and construction of the new CFES apparatus but also had to write the operation and malfunction procedures for the shuttle flight data file. Apart from that, he had to go through NASA's foreshortened physical and psychological screening for payload specialists — which consisted of a brief physical exam, an interview with a NASA contract psychologist, and an interlude in the rescue sphere. His preflight training included 125 hours of workbook instruction and basic hands-on functions, including use of the food system, the toilet, the cameras; some time was spent in the KC-135 and the T-38. Simulations with the STS 41-D crew, both in the 1-g trainer and the various simulators, began approximately six months before launch with intensive simulation two months before launch.

The decision to fly Walker turned out to be the correct one. The equipment turned out to be much more troublesome than expected during the September 1984 flight, and required all of Walker's expertise and attention to keep it going.

The future will hold more flights for Walker, an unusual circumstance for any payload specialist. He will fly at least four more times in connection with the EDS projects.

Although Charles Walker's dedication, expertise, and good timing got him a ticket into space, not all industries are as quick as pharmaceuticals to pick up on space basing part of their operations.

Microgravity Associates and John Deere Inc., two more companies that have joint business agreements with NASA, have less certain markets and somewhat different technological problems that Johnson & Johnson and McDonnell-Douglas did.

Microgravity is made up of a retired Air Force officer, a retired NASA manager (both with experience in dealing with government contracts), and a Harvard business school MBA. Backed by 12 wealthy individuals with experience in very high-risk ventures, the Microgravity entrepreneurs hope to contract an aerospace company to build a powerful furnace in which large crystals can be grown. On both Skylab and Apollo/Soyuz, large crystals were grown. The equipment to grow such crystals on a large scale and a regular basis remains to be built.

The first Microgravity experiment is a furnace that will seek out very basic information and answer fundamental questions about crystal growth in space: if you grow a crystal of commercial size and you grow it for several days, do you have more defects and do you have any more problems as a result of the extra time it has to be in the furnace to grow? Do you have to develop a process or a new growth medium in trying to develop a crystal an inch across rather than a millimeter across?

However, the most important question Microgravity will have to answer is: is there a market for space-grown crystals? Potentially, large crystals can overcome the limitations of the silicon chip in enhancing or even creating artificial intelligence in computers. Large crystals are essentially grown and then cut up in thin slices, allowing a large amount of information to be placed on each surface. "At first people with large space systems, large defense systems or very large and high speed computers will buy such crystals, "said Russell Ramsland, the Harvard-educated executive vice-president of Microgravity. "The biggest promise is to get crystals big enough for artificial intelligence — huge memory capabilities and speed — everything you'd like to get with artificial intelligence. Robotics can benefit too. Anything that has the capability to learn from its mistakes."

Although the promise is there, the market isn't, not yet. Therefore, on a series of test flights starting in late 1986, the crystal-growing technology will be tested, the crystals grown and then test marketed. At the end of the six test flights, spaced approximately six months apart, Microgravity should know whether its product will sell or not and whether the technology to support the crystal growth is reliable.

"We're essentially making the first Xerox machine," Ramsland said. "They always said you'd never have more than a seven or eight hundred machines a year market back when Xerox started, because people could use carbon paper and so on. They never thought the industry could spawn on itself."

Ramsland thought it unlikely that a technician will be flown with the crystal growth equipment at first. Since it is very large and consumes a great deal of power, the furnace will be located in the payload bay of the shuttle. It is designed to be semiautomatic, where an astronaut may merely have to turn the furnace on and off and monitor it occasionally. However, it's possible that a crystal growth expert may be needed for special crystal growth furnaces that require monitoring, observations, and recommendations for future design improvements.

There are many unknowns in this business venture that frankly scare investors off. And it takes a special investor to be able to buy into such a business. "Our backers," said Ramsland, "are a cross-section of wealthy individuals who are familiar with high-risk ventures. Some are space oriented, some aren't. Some are interested in this venture because of the large upside if it works. Others want to be part of the new frontier. Some are looking at it for their kids. I don't think an investor gets into this for any single reason. I think there might be a romance side to it too. Because of securities laws, small investors are kept out of this kind of venture, and we're a little far removed from traditional venture capital financing yet. For space business, venture capital brokers have to alter their time frame."

Skylab showed that metallurgy might benefit from the space environment as well. The John Deere Company had one of the first agreements with NASA. One of the earliest experiments involved melting iron in a crucible furnace, in order to test whether convection or diffusion was the primary factor influencing the process. They flew the experiment aboard the KC-135 and found that indeed diffusion was the primary process. Spaceflights were never done, because they were unnecessary.

However, other experiments, done by John Deere in conjunction with the University of Alabama, are testing metals potentials in other areas. These potentials may render stronger substances to be used in certain failure points in machinery, extending their lives and saving industry retooling expenses by using strong alloys in the machinery. Conductive fibers might also become a space industry.

The problem with metallurgical processing in space is that metals are heavy and require large launch expenses. They are bulky and some of the processes are, currently, labor intensive. Dr. Gary Workman, a physical chemist for the John Deere Company, looked beyond the next 20 years, when metallurgy and manufacturing of strong metals might be done from the asteroids. "It would be a one-way trip," Workman said. "Robots could gather the resources right there, process them and ship them back."

There is also money in providing facilities in space, a place where space manufacturers, like McDonnell-Douglas and Microgravity, can park their modules. One such space platform is called "Lease-craft," to be built by Fairchild and launched in 1987. It is simply a large platform that can provide attitude control and a stable orbit for its customers. It is conceivable that a Fairchild payload specialist could oversee its launch and positioning.

Beyond that, Space Industries Incorporated, a Houston-based company, has plans to launch a minimal space station. Visited occasionally by the shuttle, the ministation, referred to as the "space industrial facility," could provide the customer with a protected environment. The facility could provide protection from the vacuum of space, thermal and radiation protection, and a place where human beings could maintain and repair equipment in a shirt-sleeve environment. It would have its own basic life support system, including air and water. The facility is attractive in many respects, since it is difficult to adjust or repair modules on unmanned platforms, in a spacesuit. The launch, positioning, and maintenance of the facility is sure to involve at least one industrial payload specialist in the late 80s and perhaps more later.

Corporate astronauts will become more numerous as more companies get involved in research and development of space processes on the U.S. space station, to be launched in the early 1990s. Until we go back to the moon, business opportunities in space will be limited to industries that produce expensive, small-volume items: drugs, electronic components, crystals, and possibly fiber optics and conductive fibers.

And it is not only from the manufacturing sector that industrial payload specialists will come. NASA's most recent policy reflects a greater willingness to fly corporate payload specialists from many kinds of industries. Corporations flying "major payloads" (loosely defined as payloads that will fly in the payload bay — like satellites — not small experiments that will fly on the middeck or as getaway specials) will be allowed to send a payload specialist into space to oversee the project. This means RCA and a host of other companies launching communications satellites and so on will be allowed to fly some person from the company in space, if they are willing to pay the hefty bill for the training and if the individual can pass all the medical and psychological tests.

Large labor-intensive projects involving a large population of spaceworkers still seem to remain somewhere beyond the year 2000. Early in the 1970s, space colonization groups such as the L-5 Society, promoted the construction of a huge space solar power satellite as the means by which large numbers of people could be allowed to work and live in space. Indeed, the space solar power satellite would be a huge project requiring at least 1,000 space construction workers and all the support industries—caterers, laundries, communications workers, and so on—that any such industry would require. Gigantic space habitats would have to be built to house these people. Space colonist enthusiasts believed that such a project was the only means by which average people — the workers, the welders, the riveters and so on—would get into space.

But the spaceman as construction worker is still a dream. While the space solar power satellite received serious study and attention, it was clearly a project before its time. It would be enormously, prohibitively expensive; would rely on habitat and support facilities that do not exist; would be costly to build; would depend on new technologies; and would have an unknown impact on the Earth's environment. In the meantime, cheaper rival power systems could challenge the SSPS. Huge mirrors orbiting the Earth, one called SOLARES, could beam light down to huge collectors on Earth — and require only launch and deployment from the shuttle, a known capability. Other plans call for building power collectors on the moon. The moon has its own raw materials which would cut down on what has to be transported from Earth, and provides a more stable platform than geosynchronous orbit does.

It is possible that one of these large engineering projects may blossom unexpectedly into a full-fledged space industry. But even without such projects, corporate astronauts may represent a potentially large number of spacefarers by the late 1990s. Businesses — some as yet unidentified — are expected to become the drivers of space development, providing the motive, means, and opportunities for true space colonization. Already business has provided a leading motive for construction of space stations on the part of both the private and government sectors. This demand is expected to grow, along with the number of corporate astronauts who will build and oversee the equipment, develop proprietary processes, and produce new products for export to Earth.

10

The Price You Pay
(Astromedicine)

Early in the space program, some doctors were convinced that man's first tentative steps into space would be tragic. The human body, after all, evolved in 1-g gravity beneath a friendly cloak of atmospheric gases. One doctor predicted gloomily that in zero gravity the heart would pump wildly until the blood vessels burst. Another expected astronauts to drown in their own vomit or starve when their food refused to pass through their gravity-dependent digestive systems. Still another was afraid that a primal fear of falling, stimulated by weightlessness, might jerk astronauts back to bleary-eyed wakefulness the moment they dozed off — and eventually drive them mad.

In fact, the list of predicted effects of weightlessness — all with some scientific basis — was enough to heighten any doctor's anxieties. They included anorexia, nausea, disorientation, sleepiness, sleeplessness, fatigue, restlessness, euphoria, hallucinations, decreased tolerance to gravity, gastrointestinal disturbance, urinary retention, diuresis, muscular incoordination, muscle atrophy, demineralization of bones, renal calculi, motion sickness, pulmonary atelectasis, tachycardia, hypertension, hypotension, cardiac arrhythmias, postflight syncope, decreased exercise capacity, reduced blood volume, reduced plasma volume, dehydration, weight loss, and infectious illness. (Nicogossian and Parker 1982: 5).

The Russians were so concerned about pilot disorientation that they designed their spacecraft to be wholly automatic, wholly controlled from the ground. Although Yuri Gagarin, the first man in orbit, maintained excellent psychological and physical health during his short one-orbit *Vostok 1* flight, the cautious Russians still prefer to guide their craft from the ground. Space motion sickness appeared for the first time during the *Vostok 2* flight of Major Gherman Titov four months after Gagarin's flight and resulted in a strong emphasis being put on testing the vestibular system of prospective cosmonauts. For instance, all potential cosmonauts were made to endure rigorous g forces in a punishing centrifuge. There was a case in which the centrifuge did so much permanent physical damage to a cosmonaut candidate that he was disqualified from flying as a cosmonaut as a result of it.

The six flights of America's Mercury astronauts succeeded in dispelling some medical fears and identifying some health problems: weight loss, dehydration, and slight cardiovascular impairment as a result of living briefly in weightlessness. In the slightly longer Gemini flights, physicians found that the astronauts lost some bone calcium, some red blood cell mass, and some muscle capacity and experienced postflight orthostatic intolerance; that is, they had difficulty readapting to Earth gravity. One surprising revelation was the great exertion and high metabolic rate required for spacewalking.

By the same token, Russian flights had similar results in the Voskhod and early Soyuz programs. One surprise that came out of the Russian space program occurred after *Soyuz 9*, a flight in which the cosmonauts had to exercise two hours a day with chest expanders, elastic tension straps, and isometric exercise devices. In spite of the exercise, the cosmonauts were so deconditioned that they had to be carried from the spacecraft after landing. Although this turned out to be an anomalous experience, the Soviets gave very careful study to what kind of exercises would benefit people the most and then instituted some very rigorous exercise regimens afterward.

During the Apollo program, physicians continued to observe decreased red cell mass and plasma volume, some muscular deconditioning, dehydration, weight loss, and some motion sickness. Also, the astronauts seemed to eat less than they really should have, a phenomenon that still occurs to many astronauts to this day.

Astronaut James Irwin on *Apollo 15* also experienced cardiac arrhythmias, which was believed to be brought on by an electrolyte imbalance, easily remedied by balancing fluids and minerals. It may turn out to have been an early symptom of his subsequent heart disease, since he had an almost fatal heart attack a few years later. Irwin's heart disease was not a result of his spaceflight, since no astronaut before or since has ever gotten such heart trouble.

Working on the moon brought no surprises. Physicians already knew that extravehicular space work involves a great deal of physical exertion. None of the astronauts experienced any trouble acclimating to the moon's small gravity forces after being weightless for three days. The astronauts noted that vision was briefly affected owing to the way light scatters on the moon. Because the moon has no atmosphere, light does not scatter as it does on Earth, where clouds, humidity, and smog all absorb light. Instead, areas that are directly lighted by the sun are very bright, while those not illuminated directly are in deep shadow. The astronauts adapted to this reordering of normal visual relationships quickly, and there were no real problems.

Then there are the mysterious "light flashes." Astronaut Buzz Aldrin reported seeing them during his trip to the moon, and they were at first believed to be a rare occurrence. Afterward, many shuttle astronauts reported the same phenomenon. "It was as if somebody had taken a picture with a flash," said astronaut Dr. Norman Thagard. According to astronaut Dr. William Thornton: "I was sleeping in the airlock and I thought something was reflecting in. Finally, one came in and reminded me of a cloud chamber picture, a burst with a couple streaks out of it." The light is believed to be cosmic radiation, which may trigger a few retinal cells in the process of traveling through the cabin and the astronaut's eyeball.

The American Skylab missions and the Russian Salyut missions were all long-duration spaceflights, allowing physicians to ponder health issues related to long-term space living. For instance, habitability concerns such as noise, waste management, life support, crew wake-sleep cycles, and so on could all be examined in detail. There has been greater opportunity to study the effects of long-term space living. Both Russian and American spaceflights proved beyond doubt that cardiovascular deconditioning and loss of calcium from the weight-bearing bones of the body were the principal dangers to good health in long-term spaceflight.

Two decades of spaceflight have dispelled some anxieties and verified others. Of the 30 predicted effects, 11 are serious enough to require countermeasures. Sometimes the effects are related and the countermeasure is simple. In others, the countermeasures to space health problems elude physicians to this day. But in these years, physicians have had a chance to observe medical problems in flight; study subtle, mysterious adaptive processes; and predict possible health hazards in the future. They now have a fairly clear scenario of human adaptation to weightlessness.

When first launched into space, the astronaut's body senses the new weightless environment almost instantly, and physiological changes occur from the first moment of achieving orbit. Body fluids redistribute themselves toward the head. As they do, the face puffs out, and the proportions of the body change, too. As the semifluid intestines float upward, an astronaut could lose as much as four inches in girth, and as his spinal curvature straightens out, he could gain an inch or two in height. Indeed, one astronaut once noted gleefully that in space he was finally taller than his wife. The straightening of the back is attended by a slight backache, which apparently all astronauts experience to a certain degree. The natural human posture in space is a somewhat curled fetal-like position, making chairs and other Earth-like furniture unnecessary.

"Backache is caused by mild stresses on the muscle," said Thornton. "The one area people are sensitive to is the back. This varies from a little low back discomfort to tension from an abnormal direction. I felt it in the superior portion of the pelvis where the muscles meet. I could notice it if I thought about it. There was usually a line of people waiting to use the treadmill the next morning, which relieved it temporarily, like getting up and stretching in the morning."

The sensitive organs of balance, the otolith and the semicircular canals of the inner ear, are part of the vestibular system, the first body system to be profoundly affected by zero gravity. These organs sense linear and angular acceleration and use visual information to help a person orient himself or herself. Because these sensitive systems have evolved in the presence of Earth's gravity, they can become addled easily. Any time the vestibular system is affected, motion sickness can result. Symptoms consist of pallor, cold sweating, mild dizziness, "stomach awareness," nausea, and vomiting. They can occur as early as an hour after launch or as late as the second day of spaceflight. Symptoms can persist for as long as four days. Astronauts have noted that rapid head and body movements aggravate symptoms. The symptoms disappear, never to return, which indicates that it takes individual vestibular systems a certain amount of time to adjust to the weightless home. Also, it appears as though veteran astronauts are less likely to get space adaptation syndrome (SAS, the term used for space motion sickness) than rookies, even if they experienced SAS on their maiden voyages. Jack Lousma, for instance, was very ill on *Skylab 3* but was less so on his flight of STS-3.

It is impossible to predict who will get motion sickness in space and who won't. Over the years, vertical oscillators, roll and pitch rockers, aerobatic T-38 flights, and parabolic flights in large KC-135's have been used to predict potential illness, all without avail. According to a NASA report on SAS, a motion experience questionnaire was handed out to veteran astronauts. "The questionnaire revealed that a few crewmen had experienced some motion sickness symptomatology during past exposure to aerobatic flight, parabolic flight, and heavy sea conditions," the report stated. "The questionnaire results did not correlate with the actual incidence of space sickness symptomatology." (SAS fact sheet, NASA release no. 83-024, 1983). In essence, motion sickness on the ground did not predict space sickness; 50 percent of astronauts get SAS, 50 percent do not. Nobody knows why. Whereas vestibular sensitivity has been used as a cosmonaut selection criteria in the Russian program, it has not been so in the United States. "I certainly would never throw a space-age Lord Nelson out of the astronaut corps because of it," said Bill Thornton.

The causes of SAS have been under intensive research in recent years, especially since it is inconvenient to have motion-sick astronauts on space shuttle flights, which are characteristically short and have excessive work loads. For years, three theories sought to account for SAS.

The one favored by the Russians is the fluid shift theory, which claims that fluid shifts produce changes in cranial pressure, pressing on certain blood vessels essential to the vestibular receptors. Many of their efforts have been devoted to counteracting the effects of the fluid shifts.

A leading theory in this country is the sensory conflict theory; that is, information from the eyes, the semicircular canals, and the otoliths are at variance with previously stored neural information and each other, resulting in motion sickness. Such a mechanism may contribute to seasickness. Yet, "After my flight, I am convinced that this is a completely different phenomenon from any one-gravity kind of motion sickness," said Thornton, an aerospace medicine expert sent up on STS-8 for the purpose of studying SAS.

Thornton believed that SAS was caused by the stomach which ceased to perform its normal functions in zero gravity. "The gut shuts down," he said. It could explain why some SAS victims experience vomiting without nausea. He recommended that Reglan, an anti-vomiting drug, be used in future missions.

Disorientation resulting from things not being where one expects them to be was thought, by astronaut Joe Allen, to play some part in SAS, but Thornton claimed that, "I'm totally unimpressed with that disorientation aspect, which had been quite high on my list prior to my flight."

This theory goes hand in hand with another called the otolith asymmetry theory. Imbalances in the otolith, the organ in the ear that senses linear acceleration, accounts for motion sickness, a contention that would seem to be supported by rigorous Spacelab 1 experiments in which visual orientation and angular and rotational acceleration were all tested, suggesting the involvement of the otolith in SAS.

Whatever the cause, treatments for SAS have been wide ranging over the years. They include parabolic flights in KC-135s to accustom the astronaut to short-term zero gravity, aerobatic maneuvers in T-38s (which Jack Lousma felt helped a great deal for his STS-3 flight), use of a rotating chair or rotating room to accustom the astronauts to the standard body movements in a disorienting environment. The effectiveness of these countermeasures are dubious at best. Astronauts Gerald Carr, William Pogue, and Edward Gibson of Skylab 4 tried to take these precautions before their flight. Two of the three

crewmembers experienced SAS regardless of these countermeasures, and one even had symptoms persisting into the fourth day of
the 84-day flight.

The Russians used decongestants briefly to alleviate fluid redistribution (which causes a nasal and sinus stuffiness that feels much
like a bad head cold). But drugs have unpredictable side effects in
space. For instance, Russian cosmonaut Vladimir Shatalov took a
decongestant and spent the rest of his day in near slumber. Drowsiness, a side effect of both decongestants and motion sickness
drugs, is a completely undesirable side effect in space, where an
impact of reduced alertness and slowed reflexes could endanger
life.

The drug most favored by the Americans was called scop/dex, a
compound of the depresssant scopolomine and the stimulant dexadrine. Both drugs have bothersome side effects — scopolomine can
cause drowsiness, and dexadrine can cause nervousness. Together
it was hoped that they would counteract each other and cure SAS.
Medical protocol changed in the early program. At first, rookie
astronauts or veteran astronauts with a history of suffering SAS were
asked to take a scop/dex pill one hour before launch. Unfortunately,
although the drug would be in the system at the time of launch
before onset of symptoms, it was deemed somewhat risky to invite
efficiency-reducing side effects during launch, the most dangerous
moments of flight. Afterwards, it was deemed worthwhile to ingest
scop/dex upon achieving orbit, but by then, the symptoms were
already present, and absorption of a drug through a potentially
upset stomach was not deemed effective either.

The Transderm V, a skin patch that guaranteed proper absorption
of the drug through the skin (bypassing the upset gastrointestinal
tract altogether) in a time-released manner, proved to be equally
ineffective. One astronaut who changed Transderm pads more frequently than he should have in order to get relief from his symptoms
actually became somewhat lightheaded and whoozy. The Transderm side effects can also include drowsiness, blurred vision, and
mouth dryness, none of which are particularly desirable. It was
dropped from space medical protocol and replaced with the old
orally administered scop/dex.

The dosage of scop/dex must be carefully adjusted for each

astronaut, because people have widely varying sensitivities to it. Sally Ride felt drowsy from her early encounters with scop/dex on the ground, whereas her commander, Bob Crippen, had the opposite response and felt so shaky that he accidentally spilled coffee on the shuttle console during a simulation, shorting out the circuits.

"I've never been a fan of scop/dex," said Thornton, "although using some other drug might be useful."

One Russian physician advocated the use of a vasodilator called cavinton for the problem. "Cavinton enhances capillary efficiency in the brain," said Dr. E. T. Matsnyev of the Biomedical Center in Moscow. "Blood flow to the cerebral cortex and the thalamus is improved." In various ground tests, cavinton did not cause deterioration of motor functions or attention span, unlike scopolomine, the drug the American space doctors use, which can cause grogginess.

In ground tests that simulate space conditions, both normal people and people prone to motion sickness were helped by using cavinton. "Subjects who were given cavinton seven days before and all during the five test days showed increased tolerance to motion sickness," said Matsnyev.

The Soviet physican warned, however, that cavinton must undergo testing in space before any claims could be made for its effectiveness there.

One NASA doctor, Patricia Cowings of Ames Research Center, suggested trying biofeedback. However, the astronauts were unwilling to undergo nearly 100 hours of biofeedback training to try it. Biofeedback training has had some success in an Air Force attempt to use it for restoring some motion-sick pilots to flight status. However, some doctors feel that biofeedback results vary from individual to individual and have to be reinforced from time to time. Its effectiveness may be tested on the dedicated biomedical flight of Spacelab 4 in early 1986.

The Russians' latest attempt includes a "neck pneumatic shock absorber," a device that controls the weight load on certain cervical vertebrae and neck muscles and restricts head movements. The device looks like an intensely uncomfortable affair, consisting of a cap, two unstretchable straps in back, and two in front attached to another strap going around the waist. Flight testing on *Salyut 6* seemed to provide some benefit in controlling the symptoms of SAS, owing, Russian physicians think, to controlling neck muscles and vestibulocervical reflexes involved with the organs of balance. (Nicogossian and Parker 1982:150).

On the flight of *Soyuz 38,* specially designed shoes that put pressure on the soles of the feet were tested, under the theory that such pressure reduced motor disturbances. The shoes are not part of the regular Russian regimen.

Russian physicians also designed a "pneumatic suit" to mitigate the problem, to little effect. The suit was actually a pair of pants that fit tightly around the upper thighs, preventing a sudden massive flow of blood from the legs to the head. NASA has such a garment for reentry purposes, to keep the blood from pooling at the feet as gravity pulls the blood down again, in order to keep astronauts from passing out during reentry and landing. But fluid shift is only one problem involved in SAS; therefore there has been little effect from the use of this suit.

"SAS is not that big a problem," said Thornton. "It's been blown out of proportion. It has not hampered a mission so far." SAS is only a temporary problem. Within three to five days, all astronauts adapt to space and the symptoms go away. However, the feasibility of space colonization is tested by other, more serious problems.

Oddly enough, vision deteriorates from 5 percent to 30 percent in the first few days of flight but has had little impact on astronaut performance. In the first few hours of his flight on *Skylab 2,* even a visually oriented person like astronaut Joe Kerwin had difficulty balancing at first. Kerwin claimed that when he closed his eyes, "My first instinct was to grab hold of whatever was nearest and just hang on, lest I fall"(Nicogossian and Parker 1982:156). Kerwin still performed adequately. Since then, astronauts have been able to carry out complex visual-motor skills, using the remote manipulating arm of the shuttle with complete accuracy, indicating that visual deterioration does not play an important part in space. Russian physicians A. I. Lazarev, and S. V. Avakyan agree, writing that in spite of up to 40 percent of contrast sensitivity loss in addition to loss of other general visual capability, ". . .the vision of cosmonauts in flight is as reliable as that on Earth. This enables the extensive use of vision to carry out scientific research and control of the space vehicle under normal levels of brightness and illumination" (Nico-gossian and Parker 1982:158).

In the early 1970s, as the Russians sallied into their long-term spaceflight Salyut program and we pursued the Skylab program, disquieting changes in various biological systems were observed. The redistribution of fluid to the brain fooled the body into elimi-nating too much water in the urine. As a result, body tissues began dehydrating, causing blood plasma volume to drop by 10 percent after two weeks. Red blood cells, which carry oxygen to the tissues of the body, were diminished by 15 percent, and their shape was changed. Bone began losing calcium at an alarming rate of 0.5 percent of total body calcium per month, and in certain bones as much as 5 percent. Over a period of six months, the Russians measured an average 14 percent calcium loss from the bones, and one cosmonaut actually lost 19 percent in that time. Heart and a host of other muscles grew weaker, and with less blood to pump and no gravity to pump against, the heart shrank slightly.

Naturally the dehydration problem is easily solved by simply drinking more water. The Russians observed in their very long flights that the change in the shape of the red blood cells and the reduction in total blood plasma were probably normal adaptations to the weightless state and did not pose any real health hazard, except possibly, immediately after returning to Earth, where the diminished number of cells may have to catch up with the suddenly increased needs of 1-g living again. In the entire 120-day cycle in which the body's blood supply is completely renewed, there was no significant functional difference between the oddly shaped "space" blood cells and the normally shaped Earth blood cells.

Atrophy of the heart and muscles is not too much of a problem for the shuttle astronauts, since their flights are so short. Even so, some feel their bodies decondition almost immediately. Robert Crippen claimed that on his second day of flight on STS-1, he began to do some pressing exercises, responding to the subtle needs of his slightly deconditioning leg muscles.

In effect, exercise is completely voluntary on the part of shuttle astronauts. Anybody can tolerate a little bit of deconditioning. It's like spending a week or two in bed with a bad cold. However, the excellent treadmill designed by Dr. Thornton does give the shuttle astronauts time to exercise, especially for many of the somewhat athletic astronauts who are used to it. The treadmill is anchored, and the astronaut dons a harness, sets up the proper bungee attachments, and runs. The tautness of the cords has been carefully calculated to give the proper amount of tension, so that legs and calf muscles get exercised without getting overfatigued. All astronauts who have used it expressed satisfaction: they felt as if they had gotten a good workout. Some astronauts claimed that twenty minutes on the treadmill reduced the stuffiness of the sinuses and even had a psychological bonus of reducing tension after a hard day and improving an astronauts' outlook by the subsequent relaxation.

"It's a noisy sucker, though," said Thornton, who invented it many years ago and used to keep it in the astronaut's coffee room closet before the shuttle flights. "We have to do something about that."

Exercise for the Russian cosmonauts on long missions is an absolute must, and they exercise for approximately three hours a day, equal, according to Dr. Oleg Gazenko, head of space medicine in the USSR, "to climbing a building 200 stories high." The exercise is believed to be the best countermeasure for the heart's adaptation to space, which includes decreased size, decreased output, and shorter contractions. Although this is normal for space, the cosmonaut becomes maladapted to Earth-living and so must exercise in order to maintain his future health on Earth.

The Russians use not only a treadmill but also a host of other devices. There is a device called the Tonis 2 that electrically stimulates otherwise unused muscles. The Chibis crimped trousers are a garment that the cosmonauts wear briefly; they create a partial vacuum around the legs and hips, reversing the effects of weightlessness briefly and approximating normal gravity-bound circulation patterns. Another in-flight suit called the Penguin suit (so called because the cosmonaut waddles like a penguin when wearing it on Earth) forces the cosmonaut into a pronounced fetal position if he relaxes. The cosmonaut must constantly use his muscles to counteract the forces of the suit.

An American version of Tonis 2 may be tried in future spaceflights to stimulate unused antigravity muscles of the legs. Transcutaneous electrical nerve stimulators (TENS) are used now on young people with a congenital deformity called scoliosis, which causes the spine to grow somewhat crooked. Patients with this condition place the stimulator pads on weaker muscles before going to bed so that the TENS can "exercise" or stimulate those muscles during sleep, correcting the condition without using painful braces. Sometimes TENS are marketed as a lazy-man's exercise device. "In space they could stimulate unused antigravity muscles," said Dr. Dan Woodard, a Houston physician, "where they could maintain stress on unstressed muscles and counteract both muscle deconditioning and bone calcium loss."

A bicycle ergometer was used both in Skylab and Salyut in an attempt to keep the heart and blood vessels in good shape.

Increased food intake is also necessary, according to Russian dietitians. Whereas American shuttle astronauts might breakfast on eggs and bacon, lunch on spaghetti and meatballs, and sup on barbecued beef, Russian cosmonauts might wake up to chicken with prunes. Outlining the typically Russian space diet, Dr. Alexis Sibuk claimed that the cosmonauts consume "approximately 3,000 calories a day," frequently flavoring the foods with plenty of "onions, garlic and bacon flavoring."

A typical breakfast might include chicken with prunes, pralines, and white Borodinsky bread, washed down with coffee or tea. Lunch might consist of cottage cheese with black currant puree, honey cake, and black currant juice. Dinner would start off with sauerkraut soup, followed by meat stew, brown Stolopy bread, prunes stuffed with walnuts, candied peel, and coffee. Day would end with a light supper of Rossyisky cheese, brown bread, chicken in tomato juice, and tea.

Occasional treats of fresh fruit and vegetables supplement routine space food whenever automatic Progress supply ships fly up to resupply the Salyut space station.

Although the diet might strike American space doctors as high in fiber and sweets, Dr. Sibuk said, "they had no dispepsic problems, their stools were normal, the diet preserved high work competence, and we saw no reason to change it." He confessed that after the traditional six-month-long flights, the cosmonauts complained of "being bored with the limited selections onboard."

The cosmonauts tended to eat more toward the end of the flights, in spite of the monotony of the menu, because they "also worked more," said Sibuk.

The Russians strongly believe that diet and rigorous exercise also counteract deterioration of muscles (brought on by lack of use of antigravity muscles), altered metabolism, and changes in digestion of certain nutrients.

The progressive loss of essential calcium from the bones has, so far, posed the thorniest problem of all. Physicians are not sure what causes the phenomenon. In not being stressed, some physicians believe, the bones lose a certain piezoelectric dynamic, a process that continuously stimulates stressed bone to produce more calcium. Others believe that lack of vitamin D, usually an essential vitamin to calcium synthesis gathered by sunlight or through vitamin supplements, or the inability of the intestines to absorb vitamin D, might cause the problem. Others postulated that the fluid redistribution causes a malfunction of a calcium regulator, such as the parathyroid gland.

Naturally, once back on Earth, calcium is absorbed properly again, and new calcium is reintegrated into the bones. Whether the new calcium is integrated into the bone matrix completely, such as to be "good as new," or whether it simply scars over, like injured bone, is as yet unknown. Most scientists concerned with this area of space medicine believe that some calcium loss — as much as 5 percent in the weight-bearing bones of the legs and spine — may be permanent. When test animals are placed in space for a long time and later autopsied, the final verdict on the reintergration of calcium may be rendered. So far, amount of calcium loss and gain is calculated by x-ray studies of the heel of the foot, a somewhat imprecise measurement and one that tells, not how calcium is reintegrated, but only if it is.

Oleg Gazenko believes that "bone demineralization poses the most serious health problem to long-term space missions, since it does not yet appear to be a self-limiting mechanism."

There are many approaches possible in counteracting the decalcification process. "One is to introduce more calcium into the body to allow calcium fixation in the bones. Another is to use biologically active preparations to regulate the calcium itself. The third is to develop mechanical loads on the bones to retain the calcium in the bones," said Gazenko.

Drugs such as the diphosphonates, especially one called Cl2 MDP, have yet to be space tested, although they showed some promise in osteoporosis bedrest studies at Ames Research Center.

In very long flights, the Russians noticed some changes in the immune system, which protects the body against infection. A rise in steroid hormones and damage to the T-lymphocytes have both been reported. After the 140- and 175- day mission on *Salyut 6*, in 1978-79, Dr. Anatoly Yegorov, the flight surgeon, reported a "heightened sensitization to a number of allergens." The cosmonauts showed increased vulnerability to staph and strep infections. Yegorov considered changes in the immune system to be a worry but stopped short of saying they were an outright danger. These changes might simply be very subtle forms of adaptation to space we do not as yet understand.

On the other hand, Dr. Joseph Degioanni, of the Johnson Space Center, believes that the steroid hormone and T-lymphocyte changes could merely be indications of stress. Long-term space voyages are, if nothing else, stressful. The immune system, always sensitive to other factors in the body, could be responding, not to a physical condition, but to a psychological one.

Dr. Gazenko claimed that it was "impossible to say at this stage whether the rise in steroids and the damage to T-lymphocytes is an immune response or a stress response. We are continuously trying to separate the effects of gravitation and weightlessness on the lymph system. In the lymphocyte population there are some that live a short time and some that live a long time. Those that appear and disappear faster are the majority. But 15 to 20% survive a long time. Because the lymphocytes don't carry a passport, it is difficult to tag them and to know which ones are in the process of destruction, which ones are old, and which ones are new. In general, there are changes in the function of the lymphocytes but at the present time, at the state of the art, it is difficult to say which lymphocytes you are dealing with . We're concerned about that, but we don't know to what degree it is a problem. There is no correlation between lymph breakdown and illness."

Currently Russian space doctors are trying to define norms in weightlessness for blood volume, heart activity, calcuim loss, and so on. Also special drugs, known as adaptogens, may play a part in maintaining health in space.

The principle of testing drugs preflight is identical in the United States and the Soviet Union. "The most common medicines are tested prior to flight," said Gazenko, "and we are cautious to make adjustments for potential changes in flight. However, the cosmonauts are not very keen to take medication."

Radiation affects body systems differently. The lymph tissues, bone marrow, gonads, and gastrointestinal lining are very sensitive to radiation, while lungs, skin, kidneys, eyes, and liver are only moderately affected, and the central nervous system, muscles, bones, and connective tissue are hardly sensitive at all.

Naturally, the longer the flight, the greater the exposure to radiation. When a space station is established out beyond the protective van Allen belts, on the moon and on Mars, effects of ionizing cosmic radiation (observed as flashes by Aldrin, Peterson, and others) may have long-term effects. Radiation is known to cause leukemia and chromosomal mutation, although nothing in Russian animal studies has yet suggested that mutations are inevitable. Still Gazenko warned: "While flights limited to a few individuals cannot in any way affect man's biological evolution, flights of crews representing a rather large population could theoretically lead to a biological evolution that is based on mutations."

Protection against radiation for low-Earth-orbit space stations consists of increased thickness of the skin of the spacecraft and possible use of future radiation shelters aboard space stations. It is possible that protective garments could be designed and antiradiation drugs used if the problem ever becomes serious, as it might if people inhabit geosynchronous orbit, the moon, or Mars. In test animals, it was found that protective garments over the abdominal area seemed to increase survival rate of high-radiation doses. Also certain drugs called antioxidants might provide some protection against radiation, although these drugs are not without damaging side effects of their own. Naturally, since radiation doses experienced by astronauts are well within health norms, these more strenuous methods have not been used, nor are they likely to be used until a definite need arises.

Now, of course, medical concerns for the shuttle astronauts involve mostly the treatment of SAS and any other incidental problems. Medical kits aboard current shuttle flights are no more than "spiffed up first-aid kits," according to Dr. Jim Logan, chief of flight medicine at the Johnson Space Center. They include bandages, emergency items that make minor surgery possible, and various drugs, both injectable and oral. The medicines include many off-the-shelf medicines such as Tylenol®, Benadryl®, Lomotil®, Actifed®, aspirin, Sudafed®, Mycolog® cream, Afrin® nasal spray, Neosporin®, Blistex®, Robitussin®, and throat lozenges, as well as prescription drugs such as Valium®, scop/dex, ampicillin, erythromycin, tetracycline and Cortisporin® otic solution, and many other drugs. Salt tablets are included for fluid loading; atropine, digoxin, lidocaine, and nitroglycerin for heart emergencies; codeine, Demerol®, and morphine for pain; isoproterenol for shock; Bupivacaine® for toothache; and Compazine® and Reglan® for vomiting. Apart from the major space health problems already mentioned, astronauts have treated themselves for flatulence, dermatitis, back pain, eye irritation, urinary tract infection, headache, muscle strain, diarrhea, constipation, ear and lung inflammation, insomnia, and cardiac arrhythmia from the medical kits.

The Russians reported that their traumas so far have been very small also — small cuts, scratches, bumps. According to Gazenko, "The healing process for these were the same in space as they were on the ground." Two cases of dental problems did occur in the Russian space program, after which the health protocol called for very strict oral hygiene. "There have been no problems since."

Diagnostic equipment aboard the shuttle is limited to a blood pressure cuff, disposable thermometers, ear and eye viewers, a magnifying glass, a stethoscope, and ophthalmoscope, an otoscope and speculum, a penlight, tongue depressors, and a urine test package. The Russians have much more diagonstic equipment aboard, more for research purposes than for anticipated health emergencies. These include a standard 12-lead electrocardiograph, vectorcardiograph, echocardiograph, rheocardiograph, phlebograph, plethysmograph, pneumogram, two radiometers, and calf volume and separate muscle group measurement devices, all with a microcomputer for downlink telemetry. However, the Russians do not, as yet, have surgical facilities aboard their space stations, nor do they plan to regularly carry physicians into space until a significant space population requires one, the stance taken by physicians in the United States, too.

Gazenko believes that both the United States and the Soviet Union "follow a program of preventive medicine as an effective countermeasure to health problems in space."

The major health worry now confronting the shuttle program involves the possibility of an astronaut's developing decompression sickness (the bends) during a spacewalk. The bends is caused by nitrogen bubbles forming and migrating through body tissues as a diver or astronaut moves between varying atmospheric pressures. During the Apollo program, a cabin pressure of five psi (pounds per square inch) was maintained with a pure oxygen atmosphere. Similar pressure and atmosphere existed in the spacesuits, so that no bends resulted. However, the shuttle was designed with a 14.7 psi atmosphere, close to that on Earth at sea level, with a combined oxygen/nitrogen atmosphere. The spacesuit is only four psi with a pure oxygen atmosphere, and bends can result without precautions.

The protocol for the first spacewalk on STS-6 was to get astronauts Story Musgrave and Donald Peterson into their spacesuits three and a half hours before the spacewalk and to have them breathe pure oxygen in order to wash out some of the nitrogen in their blood. Such a procedure is time comsuming and tedious, and during ground tests resulted in 30 percent of test subjects' having some symptoms of the bends, seven percent of whom had to discontinue the tests because the symptoms were so bad. Although Peterson and Musgrave did not experience the bends themselves, the lengthy procedure is very costly in both money and time, removing two astronauts, as it does, from active work for nearly four hours per spacewalk.

Another protocol called for lowering the cabin pressure to 10.2 psi and prebreathing pure oxygen only an hour before the space-walk. There was concern for the air-cooled electronics aboard the shuttle, which could overheat and malfunction in lowered pressure. The pressure aboard STS-7 was lowered to 10.2 psi for 30 hours, and the results were good. Electronic temperatures ran below pre-diction, the crew felt no discomfort, and only certain fans had to perform at a higher rate.

As a result, the latest prebreathe protocol calls for the EVA astronauts' prebreathing oxygen for one hour and the dropping of the cabin pressure from the normal 14.7 to 10.2 psi for at least 12 hours before the EVA. Also before the EVA, the spacewalking astronauts will have to again prebreathe oxygen for 40 minutes before leaving the cabin. Only Spacelab flights will maintain the original prebreathe protocol used by Peterson and Musgrave. The major concern with the Spacelab is that the experiments and the hardware have not been certified to fly at lower atmospheric pressures.

There are many degrees to which an astronaut or a diver might suffer the bends. Symptoms range from mild pain around the joints and itchy skin to headache, blurred vision, convulsions, paralysis, vertigo, nausea, and death. Space physicians expect that if an astronaut gets the bends, it will be the mildest form of it: joint awareness, intermittent pain, at the most persistent pain that might restrict activity. Several factors can influence whether a person gets the bends: amount of body fat, metabolic rate, amount of prebreathe protection, and so on. The occurrence of limb bends symptoms could abort a spacewalk. Most physicians believe that 75 percent of the limb soreness will be relieved when the cabin pressure is raised to 14.7 psi. If the pressure fails to relieve the symptoms, the astronaut will be given fluids and pure oxygen, in addition to aspirin, whose anticlotting properties may relieve the symptoms. The flight surgeon may order use of steroids, too. The astronaut would then rest, and all symptoms should disappear within 30 or 40 minutes. If all goes well, the astronaut could participate in another spacewalk after a two-day rest, with a longer prebreathe required. If the astronaut still has some symptoms of the bends, the airlock could be used as a hyperbaric chamber, or the whole mission could be called off, the latter only if the symptoms were serious.

The Russians maintain a cabin pressure of 15 psi, and their spacesuits maintain a pressure of 6 psi. So far they have encountered no problems.

Currently astronauts at the Johnson Space Center are working with researchers at Ames Research Center on development of an 8-psi suit.

Both the American and the Russian space programs will be heavily involved in analyzing medical data from their flights in the future. Gazenko is interested in prolonging flights for medical studies. "The main question is to get more people in space to get a baseline on physical adaptation to space. When you talk about long-duration spaceflight, you talk about things other than physiological adaptation. You have to consider psychological motivation, the readiness to renege on the usual course of life and to pursue a career in unusual environments. In principle, I feel that the preparedness of the person to adapt himself plays a role."

Once space stations begin operations and people begin long-term work in space, they may get occupational injuries such as exposure to toxic substance, radiation, and burns and trauma of all types, including fractures, lacerations, penetrating wounds, and possibly decompression sickness as a result of space walking.

Dr. Bruce Houtchens of the University of Texas Health Science Center in Houston sponsored a seminar in 1983 at the Lunar and Planetary Institute near the Johnson Space Center to discuss the problems of emergency treatment and surgery in space. All the medical figures there agreed that surgical procedures have to be space tested, and means have to be developed for preventing blood and body fluids from floating up. Houtchens has designed a surgical module for meeting space station emergencies in the future. It is described as a table about four feet long, and he hopes to learn how instruments, organs, body fluids, and surgical techniques fare in zero gravity. Eventually, test animals will be used to evaluate surgical performance. He has already tested some parts of the system in brief KC-135 flights.

Further in the future, when people find it feasible to do major surgery or have babies in space, German doctor H. G. Mutke of Munich has invented a transparent plastic sack, fixed airtight around the patient, containing sterilized medical equipment and long sleeves for the operations. The sack is stored in a collapsed state and is inflated around the patient by means of electrical pump. Because it is very light and takes up very little space, the sack is attractive for spacework. Also the sack creates a sterile field around the patient while containing the mess surgery or childbirth is bound to cause zero gravity, where blood and tissue can float freely and contaminate the entire cabin.

In fact having children in space may not be too farfetched. Russian animal studies have not so far encountered mutations or problems in breeding various species of animals. Of flies, Gazenko said: "They had no problem in breeding and can continue to conquer space." Some anomalies appeared in amphibian development "but the data was limited and the problems may not be related to zero gravity." Fish did well but birds did not. "75 percent were normal but the experiment was not completed, due to a technical problem." Technical difficulties also arose with some mammalian experiments, on rats, and will be repeated later on Salyut and Biosat flights. The next thousand people in space will have a profound effect on future space medicine by providing space physicians with a large baseline of subjects from which deviations from health norms will be judged.

It is unlikely that any of the next thousand people in space will plan to stay there permanently, never to return to Earth. But it is possible in the next century that some individuals will not want to go back to Earth. When that happens the aims of space medicine will change drastically. Now, physicians have to keep the body healthy for two environments: weightless space and one-gravity Earth. When the need to keep the body healthy for return to Earth is removed, the burden of rigorous exercise and special diet may also be changed. Heart, muscle, bone, and blood will be allowed to find their own state of equilibrium then, weight-bearing legs and feet may become vestigial, proportions will change, posture will curl slightly.

According to Dr. Jaime Miquel, an Ames Research Center neurobiologist and experimental gerontologist, human beings may live longer in space. The body undergoes changes in space similar to aging on the ground: a decrease in grip strength, body weight, and muscle mass; lower cardiac output and respiratory capacity; loss of calcium from the bones. But Miquel reasons that one third of the calories ingested on Earth provide energy to counteract the effects of gravity. Contrary to Russian belief, Miquel believes that people could use fewer calories instead of more to sustain life, lowering metabolic rate in the process. A lower metabolic rate tends to slow time-dependent disorganization of the cells and organs, in essence, slowing the aging process.

Nobody inside space medicine or out of it is yet willing to accept the profound changes that the weightless environment, if left to itself, will wreak on the human form, even with the possible bonus of longer life.

11
Living on Space Stations

The space shuttle has, to a degree, determined how people will colonize space. Its 15- x 60-foot cargo bay restricts the size of objects that can be deployed and the methods used to construct future habitats.

By 1986 or 1987, unmanned platforms like Fairchild's "Leasecraft" may provide the parking facilities for unmanned industrial modules, such as McDonnell-Douglas' EOS pharmaceutical factory or Microgravity's crystal-growing furnaces. By the late 1980s, however, industry's needs for space products will far outstrip the limited unmanned facilities and demand a full space station. "Products will be brought to market much faster," said James Rose of McDonnell-Douglas, "and much more cheaply because time and money isn't wasted waiting for infrequent and short shuttle flights to provide necessary research data."

The planned space station is supposed to meet these needs. It could support a much-needed industry — that of repair and maintenance of faulty satellites both in low-Earth orbit and geosynchronous orbit, saving many billions of dollars throughout the economy. It would be convenient to store reusable upper stages of rockets and excess propellants from the space shuttle orbiters, both of which are tossed overboard now.

With a keen eye toward economy, Boeing, Rockwell, and other aerospace companies have designed a space station with a minimum number of costly space shuttle launches and maximum use of the humble accommodations. According to the earliest designs, the first section to be launched would be a short, cylindrical structure housing batteries, propellant tanks, and internal power equipment. Two solar arrays would be deployed from it, as well as assorted antennae. A second launch would bring up the first command module, which would house the basics of human life: communications, life support, control and stabilization equipment, and plenty of ports and airlocks for arriving and departing shuttle flights and space tugs and for additional habitation modules. With just two launches, the United States would have a little outpost in space, manned continuously by rotating crews of two. With five, it could have a thriving research and manufacturing center supporting at least eight crewmembers. Two of those three additional launches would contain, according to the early scenarios, habitation modules — crew quarters, a galley, a gym, and a medical facility. Safety features on the habitation modules include two exits each in case of fire and a storm cellar of about 8 by 15 feet in which the crew can huddle during a solar storm. The Lockheed study proposes doing away with the storm cellar altogether because avionics equipment is an excellent protection from radiation. Said Marianne Rudisill, a human factors engineer for Lockheed: "The crew members may climb under the floors and surround themselves with avionics equipment."

Apart from the basic research facilities aboard the station, science can be supplemented by using specialized modules, which can be docked and undocked from the station at will for specific tasks. These modules can be built by industries, research organizations, foreign countries, or the Department of Defense and might involve tasks such as astronomy, materials processing, pharmaceuticals manufacture, or Earth observation.

The people most likely to work on this early station will be current NASA astronauts, visiting scientific and corporate payload specialists, and, possibly, military space specialists. The scientists and corporate specialists will be probing the means of understanding the basic dynamics of matter in space — how fluids, metals, gases, and biological systems function in zero gravity; they will be designing and bringing to market new space products; and, if President Reagan's predictions are correct, they may find a way to end war on Earth from space.

Habitability experts are well aware that getting human beings to live in space long term is quite a different matter from getting them up in the shuttle for a week. "A person can put up with almost anything for a week," said Rudisill, "but problems come out of the woodwork when people have been closely confined for a long time." The average tour of duty aboard the station will probably be about three months. The Russian experience in long-term space-flight serves as a kind of guideline for long-duration spaceflight. The cosmonauts found that three months in space was tolerable. Work efficiency and spirits remained high. Six months was felt to be much too long. Work efficiency dropped, tempers flared, fatigue set in. "The cosmonauts who wish to prolong spaceflight are the ones who have never flown," said Dr. Oleg Gazenko, head of Russian space medicine. "The ones who have been on a long-duration spaceflight don't like it at all."

In order to reduce psychological stress several design recommendations have already been made. Sightseeing is the key form of relaxation in space; therefore windows are highly desirable. Human factors engineers believe that astronaut Gordon Fullerton might have had a good idea when he claimed that windows should be on all four sides of a space station to get a 360-degree view of the outside. Good food is a must, too. The Skylab astronauts suggested that more filet mignons and ice cream be put aboard, since these were the most swiftly depleted. Russian cosmonauts eat a great many sweets in space, too, including various confections, cookies, and special fruit drinks. Alcohol, a known troublemaker, will be forbidden on the space station owing to the problems triggered by it. No provision is being made for sex aboard it, and strict guidelines may be laid down forbidding sexual activity for the duration of the space tour. "Antarctic experience showed that pairing raises tensions in the whole group," said Rudisill. "Besides that, it is unlikely they'll even allow married couples to fly together because they perceive themselves as a unit, and this undermines the group."

Life on the first station will be very basic. Because there won't be much water available, the crew will not have a bath or shower as earthlings know them for about three months; neither will there be water for laundry. Once advanced water recycling technology has been established, it may be possible to luxuriate in a shower with water from recycled urine.

According to Rudisill, there will be many changes in shuttle display and control technology. "There will be use of color as a code for information, so a person can look at a screen and know what's going on," she said. "Smart machines can take over some human tasks, pictorial and graphic displays may replace numbers. Portable terminals can be plugged into the wall throughout the station instead of statically anchored to one immovable display unit."

Current wisdom dictates that the most that could be housed aboard the station would be sixteen. The modular construction limits the living area within each pod, and it would be impossible to create a station requiring large facilities without paying an exorbitant amount of money in launch costs alone. Perhaps by the mid- or late 1990s, the station will already be obsolete and space industry will be turning to yet another generation of space cities. In essence, the way station will give way to the town.

One lengthy ongoing study at the University of Houston, designed what might someday become the first space town. Headed by Dr. Larry Bell, director of the College of Architecture's Environmental Center, the 200-foot-long "Spacehab," as it was called, tried to take into consideration engineering constraints, human comfort cost, and feasible construction.

Spacehab is intended to support future research and industrial projects in space, including materials processing, drug manufacture, microgravity metallurgy, Earth observation, and genetic engineering. "Special satellites too fragile to assemble and launch from Earth can be built and launched from Spacehab," says Bell. In its 20- to 30-year lifetime, Spacehab workers could oversee construction of spaceships going to Mars and the asteroids, could support mining facilities on the moon, and could become a gateway to outer space in the same way that St. Louis did for the pioneers who moved West.

At least a fifth of Spacehab's 150 residents would be support personnel — the command crew, medical staff, galley workers, housekeepers, and so on. "It's an expensive piece of real estate," explains Bell. "Worth at least a few billion dollars."

The materials from which Spacehab will be built are those already currently used in airplanes, sailboats, and houses. When used in

Spacehab, these materials will protect space workers from meteroids, solar radiation, and the vacuum of space.

The architects envision Spacehab as being a hybrid of two distinct types of space structures: the already space-tested metal tank/module and the still untested inflatable pressure structure. To pack the most living space into easily transported components, the architects decreed that the Spacehab inhabitants must live in balloons — eight mushroom-shaped, inflatable pods clustered around two cylindrical core modules assembled on Earth. The metal modular central core, containing all the electronics, life support systems, command stations, communications systems, and some propulsion would be launched first, followed by the inflatable habitation pods.

The habitation pod walls would be made up of amazing Kelvar 49 mesh fabric, "as thin as paper and five times stronger than steel," said Guillermo Trotti, a Houston architect involved in designing Spacehab. "It would provide the structure and protect inhabitants from micrometeors. But because Kevlar is sensitive to ultraviolet, metallic mylar with a thin layer of aluminum would be glued onto the outside of the Kevlar."

Kevlar 49 is a protean Dacron-like fabric that can be formed into an airplane wing, a rope — or a space station. Its first space test may come 1987, when NASA proposes to use a 60-mile-long Kevlar cable with an attached small satellite to troll the Earth's upper atmosphere for atmospheric, magnetospheric, and gravitational data. For shuttle launch of the habitation pods, the Spacehab's Kevlar walls would be folded like an accordion and reefed like a parachute to fit into the payload bay. Because the habitation pods would measure 80 feet in diameter once inflated, high-tensile graphite ribs could be placed at intervals to keep the exterior walls taut, just as metal ribs keep the fabric of umbrellas taut. Bell explained: "High-tensile graphite is very strong and is currently used in hang gliders and the masts of boats." An additional layer of foam thermal insulation would be glued onto the Kevlar in space.

Interior walls would consist of thin layers of foam between fabric. Air conditioning and electrical lines would flow through flexible piping along the outside walls.

Additional work stations, such as for astronomical instruments too sensitive to tolerate the perturbations of Spacehab and for dangerous volatile materials, could be located in free-flying laboratories a short distance away. The free flyers could serve as emergency evacuation sites, too, since they would probably be pressurized and would have basic life support systems.

Keeping people comfortable in their alien surroundings was the architects' major concern. "Disorientation in zero gravity and isolation are the two greatest challenges to the space architect," says Trotti. The Spacehab designers took a radical approach in space station design by making human psychological needs rather than engineering expediency the driver of the interior design. The ordinary laws of bodily motion are suspended in zero gravity and human spatial orientation changes accordingly. Earth furniture is useless, liquids do not pour, objects do not stay in place unless they're anchored, and people themselves can get disoriented easily by floating into a room differently. "I remember going into compartments upside down," said Gerald Carr, the *Skylab 4* commander and adviser to the Spacehab group. "I never got lost, but I saw lots of things in different ways."

Apart from that, the psychological pressures on Spacehab residents will be similar to those experienced by Alaska pipeline workers, Antarctic explorers, and crews on submarines and oil rigs. Space work will not be an adventure like a voyage to the moon but sustained, monotonous work. "No matter how nice it is, no matter how beautiful the Earth looks from space, " said Trotti, "you can't go out of that Kevlar skin. It's a little like jail." Space workers may get nervous about near misses from space debris, which could puncture the Kevlar skin, or become alarmed by solar storm activity.

Spacehab architects have sought to overcome both the disorientation problems and the psychological stresses by mixing the familiar with the unusual in the layout of the living areas.

The showers, for instance, are a combination car wash, health club routine. The resident would first strip in a locker room and put his or her uniform in an automatic washer-dryer. Next, he or she might prefer to relax in a small sauna room or could go on to the shower room. The individual stalls would have a nozzle to spray water and sponges and vacuum devices to collect the excess water, which would bead up on the skin and on the shower surfaces. A drying room adjacent to the shower room would have blow dryers to remove more moisture, although the person might need to be anchored so as not to get buffeted around by the air jets. Sun lamps could also help dry the skin, stimulate vitamin D production in the skin, and "preserve a healthy look," said Trotti. The whole routine would take about 40 minutes, enough time for the uniform to go through a wash and dry cycle.

"The person could do this before sleep, after exercise or before work, every day or every few days," said Trotti. "The health club analogy is a good one. A person could exercise, sauna, and shower, and then have a good meal. It would feel great."

Vigorous daily exercise will be a must for Spacehab residents since the body deconditions very quickly in zero gravity. To offset muscle and heart weakening, the groups has designed a large gym with specialized equipment. Specially designed treadmills, bicycle ergometers, and rowboats would have straps and bungee cords to keep the person in place and special weights to simulate Earth stresses. Isometric equipment seen in many health clubs might also be used to build up specific muscles. An early gym design positioned equipment on the ceilings and walls for maximum utilization of space. However, exercise in space is both boring and unpleasant, and recent designs feature special lighting fixtures to give daylight and twilight effects. Wall space has been devoted to isometric skis, chinning bars, and large television displays of outdoor scenes. People on the treadmills and bicycle ergometers can contemplate projected outdoor scenes as they jog or cycle. Similarly, all isometric equipment is equipped with small digital displays so that the person can keep track of his strength and adjust weight loads as necessary.

The designers even contemplated a "scuba diving pool" located near recreation or dining room windows so that people could relax and watch the divers. However, every ounce of water would have to be recycled, including shower and laundry water, urine, and even atmospheric humidity. "We abandonned the idea as an unfeasible extravagance because it would take up 1,000 gallons of water," said Trotti.

Maynard Dalton, a NASA spacecraft designer in Houston, advised setting aside a large open area for space acrobatics and perhaps even new "zero-gravity team sports." "It doesn't do a bit of good for conditioning the body," said Dalton, "but it should be great fun." On Skylab, the astronauts frequently used a large open area for acrobatics, consequently reducing their feelings of claustrophobia and enhancing the thrills and wonder of the body's freedom of movement in weightlessness.

Work stations will have to be designed to compensate for the mistakes people are bound to make in zero gravity. Skylab astronauts found, for instance, that little items like bolts, pencils, and tools would float away, only to be retrieved later from the air uptake filters. "It might be feasible to arrange some work stations as partially enclosed bubbles," said Trotti, "each with its own ventilation filter to catch stray objects, its own light source, and its own foot restraints to anchor the worker, since chairs will be unnecessary." Vacuum suction table tops might hold papers down, or workers may prefer to use small CRT screens to take notes. Clamps, bungee cords, and magnetized tools would all be useful. Alan Bean, commander of *Skylab* 3, lined his personal hygiene kit with Velcro and wrapped Velcro around his brushes, combs, and other small items to keep them in place. The same could be done with Spacehab tool kits.

Shuttle astronauts found that they were irritated with the "jack in the box" effect in which contents would float out of tightly packed drawers every time they were opened. Bell's group designed transparent plexiglass drawers divided into numerous self-enclosed compartments to avoid this problem. "In this way a person can see what he needs before he opens the drawer to get it, and then could get the item without disturbing the other contents of the drawer," said Bell.

Because space is limited on the station, some rooms will have to adapt to more than one function. The recreation room could serve as a small auditorium in which people could see movies, listen to concerts, hear a guest speaker, or attend a large briefing. The "seats" would consist of metal frames complete with foot restraints, a back and head rest, and a hoop around the torso. The seats could be arranged in double tiers, much like a theatre with a ground floor and balcony. In order to transform the auditorium into a conference room, the seats would simply be folded back against the wall, and a conference table, with individual CRT's and foot restraints at each place, would be unfolded from a compartment in the floor.

The dining room was designed to make the most of three-dimensional space. Since there is no up and down in space, there are two dining rooms in the space of one: one right side up and the other upside down. Tables for two, four, and eight contain foot restraints, food tray holds, and an outer rail to act as a back rest and a barrier to keep diners from bumping into one another. Small windows may be located on the outside walls of the dining area so that diners might enjoy the spectacular view of the Earth passing beneath them. Although it may take some time to get used to seeing half the people in the cafeteria eating upside down, the Spacehab workers will probably learn to adapt. "One of the great things about weightlessness," said Carr, "is that you reorient quickly. Once you're accustomed to it, it's fun."

Lighting can be changed to alter the cafeteria atmosphere, beautiful views can be projected on inner walls, special foods can be served on holidays or on "ethnic evenings," and dining hours can be flexible to accommodate working schedules. Food texture blends and colorful plate arrangements might all make food more pleasing since the sense of smell is somewhat dulled in space. "Heavy use of spices may prove to be desirable," said Clinton Rappole, Dean of the Hilton School of Hotel Management and a Spacehab co-investigator. Because food in its raw state tends to occupy a large volume and weigh more at launch, dehydrated, frozen, and retort-pouched foods are all likely candidates for use in space, along with foods kept fresh by aseptic packaging and irradiation. Dispensing areas will display a wide range of foods, automat style. Each diner could select his food, rehydrate some entree items, and heat them in small convection ovens. Computers could help the cook keep inventory and plan meals.

As on the early space stations, alcohol will probably be forbidden on Spacehab too. Alcohol is routinely banned from oil rigs and other such places, since carelessness poses such an extreme hazard to crew safety. "It's not the one glass of wine that's the problem," said Rappole, "but the fact that it would have to be stored in large amounts, causing a security problem. People devise ingenious methods to get at limited liquor supplies."

Since food is such a principal morale booster on oil rigs, Antarctic voyages, and other isolated, stressful endeavors, special attention will be paid to menu variation, attractive presentation, and abundance aboard Spacehab. Some limited amount of fresh food can be flown up on routine shuttle missions, or some area may be set aside for a space garden in which some vegetables and fruits could be grown. Such an agricultural project would have to be carefully planned, however, in order to produce the most food for the least space.

University of Houston biology professor Joe Cowles (who is not a participant of the Spacehab project) learned a great deal from his plant experiment flown on the space shuttle. "We found that in mung beans, some plants were disoriented — some grew with their roots up or shoots sideways," said Cowles. "However, if the seeds were carefully positioned at planting and a support matrix were provided, the problem might be solved. In space agriculture, selection of high-producing dwarf plants, perhaps, specially genetically engineered ones, may work best. Also close attention must be paid to soil mixtures and regular crop rotation."

Some Spacehab residents may relax by working in the space garden too. "Plants seem to make people feel secure," said Cowles. "We've been in this world awhile, and whether it's a habit or a necessity, we seem to need plants psychologically. There aren't many homes that don't have plants, and it seems not to be enough just to see them. We seem to need to tend them too."

The Skylab astronauts found that they liked to spend time in the wardroom, a cozy, small space. The Spacehab architects designed the individual bedrooms with coziness in mind: the rooms measure only seven feet long and five feet high and wide. Small, transparent pouches on the wall would hold a few personal belongings, since uniforms would be provided and personal belongings limited to only ten pounds per resident. Each room has a sleeping bag that folds away and a CRT screen with fold-out writing surface. The CRT could act as an electronic mailbox, library, correspondence course classroom, telephone, window, video game, T.V., and entertainment center in one. Workers could also adjust the lighting and air conditioning in their rooms from there.

In order to keep health complications to a minimum, people with chronic diseases such as diabetes or heart disease will probably not be allowed on Spacehab. However, doctors do expect that the usual coughs, colds, and indigestion will occur. Also physicians have to be prepared to treat injuries caused by industrial accidents: fractures, toxic exposure, burns, and penetrating wounds to the chest, abdomen, or head. Since medevacing an injured worker to Earth is unlikely, the Spacehab physicians will undoubtedly be surgeons and the nurses trained anesthetists. "For special cases, communication with the ground may be necessary, so the surgeon can be talked through a very specialized surgical procedure," said Dr. Daniel Woodard, a Houston physician who consults with Bell's group regularly. "Surgery in space will require different surgical techniques too. For instance, when you open up the abdomen, the bowels will rise up. They're anchored, of course, but it might be hard to stuff them back into the abdomen after surgery. Special retractors and special clamps will have to be designed."

Similarly, heavy, bulky x-ray equipment with huge film developers will be too cumbersome. "Digital radiography is the obvious approach for space diagnostic equipment since it does away with the film developer, and the image can be displayed on a CRT screen," said Woodard. "In fact, digital line-scan radiography will theoretically work in space better than on Earth because you can rotate the patient in the x-ray beam and obtain an image similar to a CAT-scan."

On Spacehab, medical records will be stored on computer, patient beds will have restraint straps but no mattresses, and each bed will have an electronic information panel like current intensive care units to continuously monitor patients' vital signs. Surgical drapes, intravenous feed lines, and other gravity-dependent medical equipment will have to be completely redesigned. The hematology lab will have to use new methods of testing blood samples. Lab technicians may use plastic gel to separate blood cells and blood plasma. Small, compact, blood analysers will be necessary too, in place of large, gravity-dependent ones used on Earth. "NASA has developed a means by which a dozen different blood tests can be done using pellets of reagents arranged around centrifuge disks," said Woodard. "A light is shone through the disks and a computer analyzes the level of components in the sample."

Special attention was given to designing an operating arena since it must maintain a completely sterile field. "On Earth," said Woodard, "contamination always travels downward, so everything above the waist is kept sterile. In space, airflow will have to establish the direction in which the contamination will flow." The doctor will be anchored in place with foot restraints and perhaps even a frame with belts. The operating table would be adjustable and the patient moved with relation to the doctor, the opposite of Earth practices.

When not in use, patient beds could be stowable, and doctors could spend their time perfecting space surgical procedures on test animals and studying the long-term effects of weightlessness on the human body.

Bell predicts that many modifications to the initial Spacehab design might lie ahead. "When we determine what these people are doing, what kind of industries will be using Spacehab, we might have to change the design," said Bell. "Or we'd have to change our interior designs if NASA decides it's feasible to retrofit shuttle external tanks for human habitations."

"Kevlar will have to be space tested, perhaps in space hangars, before we can depend on it as a building material," said Trotti. "What's more, there may be better, lighter, stronger materials by then."

"I think Kevlar may be an ancestor to the material we may use someday in an inflatable space structure," said Dalton. "However, to make a pneumatic skin impenetrable to meteoroids, you might want to use a self-sealing material. Goodyear brought out a material made of little plastic balls about ten years ago. Air would cause the little balls to melt, so when you got a puncture, air would start to come in, melt the balls and fill up the hole."

NASA looked at the feasibility of pneumatic structures ten years ago but had a difficult time putting much confidence in such thin walls. "Pneumatic structures would have to consist of several exterior layers, and repairs would be more than just putting a tire patch over a leak," said Dalton.

"The orbital drag would be horrendous on such a large structure," said Dalton, "and things like propulsion needs, solar array electrical capabilities, life support technology and interfaces between pod and core electrical systems all have to be worked out. It is not easy to just scale up the life support and recycling systems we're developing now. It would be better to spread several smaller life support systems throughout the structure than centralize one large one in the central core."

Although futurists have always envisioned space housing to follow a linear progression from the eight-person space stations, to Spacehab-type towns of 150, to O'Neill-type space colonies of 10,000, Dalton foresees a different kind of space migration and colonization. "I think when we move into space, we'll build a number of small specialized space villages of 20 or so people," Dalton said. "When you get too much going on at a large space station, you end up with conflicting power requirements, complex redundancies, less efficient life support, and so on. I don't see the need for large space stations for a very long time. Ultimately, when we do have larger space structures, maintaining a happy environment for people will not drive the design — engineering concerns like power, propulsion, the project mission purpose will." Once space villages do become established, all agree that changes will occur in the people themselves, in how they perceive the world, and what values may become important as a result of the space experience.

"Being out there, looking back at Earth, washes away notions of nationalism and provincialism," said *Skylab 4* astronaut Gerald Carr. "You can't help feeling that man is a little insignificant because you have to watch very carefully to see anything man has done on Earth. I came back with a different attitude toward life, toward the environment and toward mankind."

"Values stressing individual liberties may give way to those which stress group goals," said Pat Musick Carr, a University of Houston psychology professor who conducted a seminar on space values. "Also I believe psychosis is a safety valve for a person under unbearable psychological pressure, so I think people need to partake in creative activities to offset stress. I call it 'preventive creativity.'"

"Like on Earth, some people in Spacehab will have hobbies, participate in sports, write poetry or play music," said Trotti. "There will be unprecedented artistic opportunities for photography and new art forms, such as, perhaps, zero-gravity glass sculpture. A few temporary visitors on Spacehab might include writers, poets, or artists. It would be good for morale. I believe that everybody who goes into space will become more creative because of a change of consciousness as a result of just being there. They'll look on Earth culture as an outsider with a different perspective, just as Westerners who go for a vacation in the Orient observe that culture from a different point of view. We'll learn about ourselves by looking at ourselves from that perspective too.

"In centuries to come, our frame of mind will be totally different because of our space skills, as different as we are now from the people of the Middle Ages. Even today we are a Renaissance culture as a result of spaceflight."

Fig. 31   Payload specialist Byron Lichtenberg, Michael Lampton, Wubbo Ockles, and Ulf Merbold walk through experiment procedures in a Spacelab 1 mockup at Marshall Spaceflight Center.   (NASA)

Fig. 32   Artist's conception of the Spacelab 1 module situated within the shuttle cargo bay. A long tunnel attaches the Spacelab module to the shuttle cabin through the airlock.   (NASA)

Fig. 33 Merbold and Lichtenberg were the first non-astronaut scientists to fly aboard an American spaceflight (STS-9). Merbold, from West Germany, was the first foreign citizen to ever fly aboard an American spacecraft. (NASA)

Fig. 34 Merbold, Lichtenberg, and mission specialist Robert Parker, work frantically to complete numerous Spacelab 1 experiments. (NASA)

Fig. 35 Solar physicists Dianne Prinz (upper left), George Simon (upper right), John-David Bartoe (lower left), and Loren Acton (lower right) are the payload specialists on Spacelab 2. (NASA)

Fig. 36 Materials scientists Taylor Wang (upper left), Mary Helen Johnston (upper right), Eugene Trinh (lower left), and Lodewijk van den Berg (lower right) are the payload specialists on Spacelab 3. (NASA)

Fig. 37 (From left to right) Shannon Lucid, Rhea Seddon, Kathryn Sullivan, Judy Resnik, Anna Fisher, and Sally Ride were the first six women admitted into the astronaut corps in 1978. (NASA)

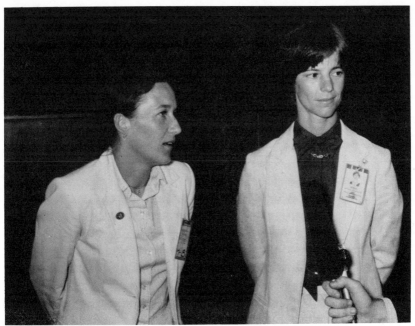

Fig. 38 Mary Cleave and Bonnie Dunbar were turned down for the astronaut corps in 1978 but made it in 1980. (NASA)

Fig. 39 Astronaut Shannon Lucid shows her exhilaration as she becomes weightless aboard a KC-135 aircraft. (NASA)

Fig. 40 Husband-and-wife astronauts Anna and Bill Fisher clown while weightless in a KC-135 training mission. (NASA)

Fig. 41 Sally Ride, America's first woman in space, monitors the continuous flow electrophoresis system (CFES) aboard STS-7. (NASA)

Fig. 42 Mission specialist Guion Bluford, America's first black astronaut to fly in space, checks a control knob on the CFES experiment aboard STS-8.   (NASA)

Fig. 43 McDonnell-Douglas's Charles Walker was the first member of private industry to fly in space. He designed and built the CFES equipment that he flew with aboard the shuttle.   (NASA)

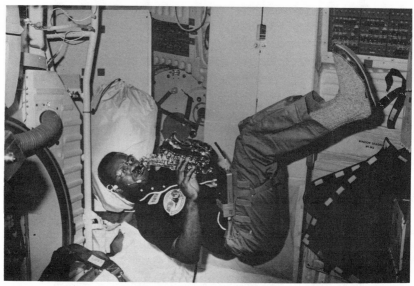
Fig. 44 Mission specialist Ron McNair plays a jazz solo on his saxophone during a spaceflight. (NASA)

Fig. 45 Ellison Onizuka, the first Asian-American astronaut, smiles during parachute practice. (NASA)

Fig. 46 The first Hispanic-American astronaut, Franklin Chang, wanted to fly in space since he was seven years old. Deeply committed to space science, Chang designs advanced propulsion systems in his spare time.　　　(NASA)

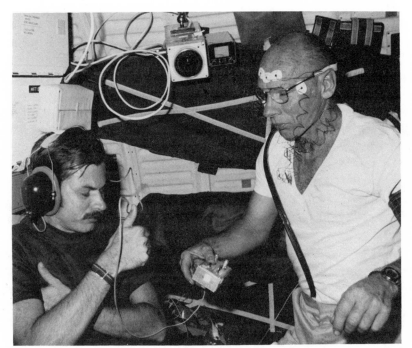

Fig. 47 Astronaut-physician William Thornton and guinea pig astronaut Dale Gardner, both mission specialists, test hearing in space during the STS-8 flight.
(NASA)

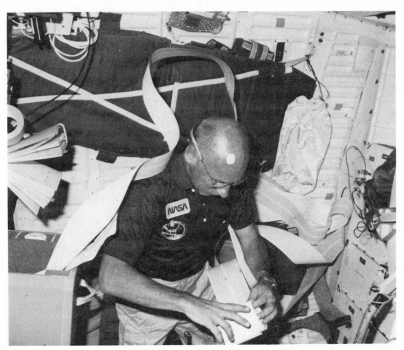

Fig. 48 Thornton checks his voluminous biomedical data readout aboard STS-8.
(NASA)

Fig. 49 How to build a space station. First come the solar arrays and the hub. (NASA)

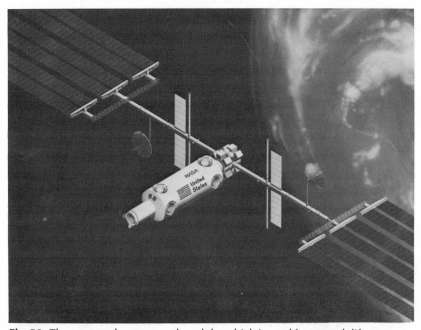

Fig. 50 Then comes the command module, which is used for control, life support, and communications. (NASA)

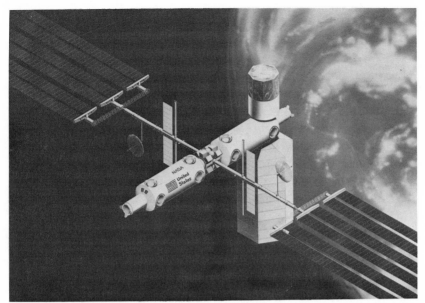

Fig. 51 Then comes a second module and a space hangar to house satellites for
(NASA)

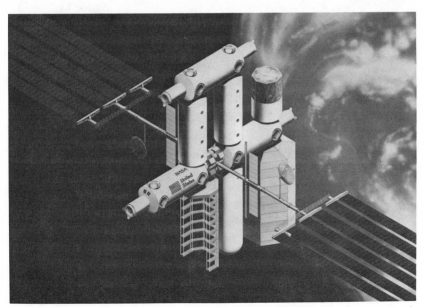

Fig. 52 Additional modules, perhaps built by foreign allies, dock to the command
modules and supply habitation and workshop facilities. (NASA)

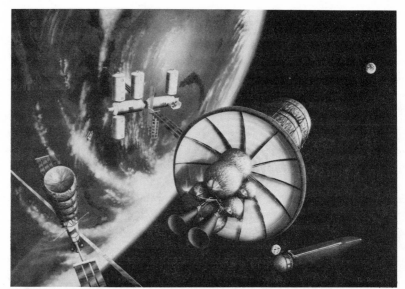

Fig. 53 An artist conceives of a large lunar ferry taking off from space station for the moon. (NASA)

Fig. 54 A lunar mining operation, overseen by human and aided by robots, is a possibility. (NASA)

Fig. 55 A lunar base may be a reality by 2010. (NASA)

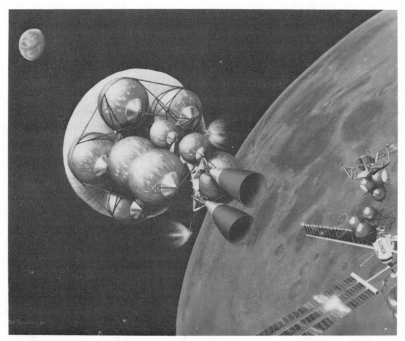
Fig. 56 A lunar ferry prepares to burn out of lunar orbit for the return trip to an Earth-orbiting space station. (NASA)

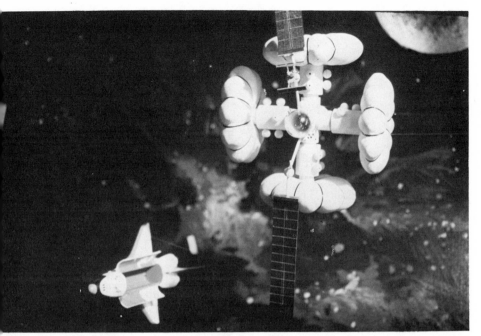

Fig. 57 Spacehab is a futuristic design for a space city whose habitation pods are reefed like parachutes for launch and then inflated like balloons in space.

(Guillermo Trotti)

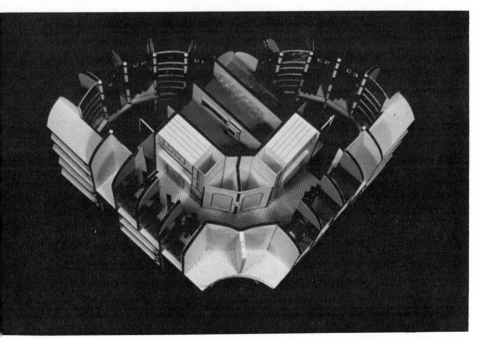

Fig. 58 Cutaway view of Spacehab's crew quarters. Bedrooms are crowded but comfortable, with black soundproof privacy curtains. Vending machines and bathrooms are arranged in the middle, for convenience.　　　(Guillermo Trotti)

12

Interplanetary Travelers

Returning to the moon has been the great dream of planetary scientists ever since the last Americans blasted off from its surface a dozen years ago. Many people believe that by the turn of the century, after the United States has built its space station, the next logical step would be to go back to the moon, this time to "turn our giant step for mankind into a giant foothold for mankind," as George Mueller, former NASA Deputy Administrator and vice president of Burroughs Corporation, said recently.

Many scientists believe that there are plenty of good, cost-effective reasons to go back, some of which focus on the moon's single great resource: oxygen.

Oxygen is, of course, an abundant resource on Earth too. In the space program, oxygen is used as a liquid propellant, especially for rockets carrying satellites into geosynchronous orbit. Unfortunately, liquid oxygen propellant is very heavy, and it costs the space users a great deal of money to lift it off the surface of the Earth into low Earth orbit. According to Hubert Davis of Eagle Engineering: "56 percent of the STS-40 payload [the Galileo probe devoted to exploring Jupiter] is liquid oxygen propellant." If moon-derived oxygen propellant were used in these satellites, satellite companies would save money on launch costs, and half of the shuttle cargo bay would be liberated for some other use. "It would be like getting two shuttle Orbiters for the price of one," said one scientist.

Davis foresees "a marginally favorable payback" in a turn-of-the-century mining operation on the moon in which bulldozer collectors, conveyers, a propellant assembly building, large storage facilities in Earth and lunar orbit, and space ferries could provide liquid oxygen propellant to future space stations, orbital transfer vehicles transporting payloads from low Earth orbit to geosynchronous orbit, and deep space probes. "A plant on the moon could process nearly 300 tons of raw lunar material per day," said Davis. "Hydrogen would be imported from Earth to be used in the propellant manufacturing process unless a lunar of source of hydrogen, perhaps in the moon's regolith, were found. Future technology could enhance the economic feasibility of such a mining operation. Of course, we need to study automation possibilities and the use of mass drivers [giant catapultlike devices]," said Davis.

The moon's aluminum, titanium, and iron could prove useful in construction of human habitats and factory machinery once human beings do settle on the moon.

Larry Haskin of the Washington University Department of Geosciences believes that moon dirt might be paydirt. "Economical use of lunar matter might involve only a single product," he said, "unprocessed lunar soil to provide shielding against solar flares for personnel and equipment on space stations." Moon dirt can be used as insulation around delicate electronics in unmanned satellites as well. Solar flares sometimes cause electrical failures and anomalies in satellites. Insulating satellites with lunar dirt may double the satellites' useful lifespan, a much-desired result for expensive satellites.

Energy is the next best candidate for lunar industry. "I'd rather get power from 20 or 30 people on the moon," said engineering consultant and lunar enthusiast David Criswell, "than have to station 100,000 soldiers around the Straits of Hormuz to guarantee oil supplies."

Criswell unveiled a plan to provide the United States with 20 gigawatts of power — enough to light up Houston and Los Angeles — from moon-based solar collectors. The preliminary moon base could grow "exponentially" to provide most United States power needs – approximately 600 gigawatts, "very quickly."

The solar collectors described by Criswell would not be the flat types we're used to seeing but "mounds of lunar dirt." Flat solar collectors would "not be very efficient on the moon," because of the relative movement of the sun. A cone-shaped collector catches rays all day long.

Several mounds would be located together in plots like a garden

(Criswell refers to them as "little power plots"). Connected together by underground wires, the solar collectors would transmit their power to a central transmitter that would convert the power to microwaves and bounce them off football field-sized antennas. From there the power could be relayed to receiving stations in low Earth orbit, to large Earth receivers in deserts and oceans, and to smaller ones near populated cities.

Building the mounds on the moon would be easy with a specially equipped lunar rover. First the rover would go over the lunar soil with a magnet, attracting little globules of iron that lie close to the surface. Then it would compact soil into mounds, lay wires between the mounds, and spread a glass film to stabilize the mound from collapse. The mound would then be covered with a sheet of iron "and you have a solar cell," said Criswell. "Both the glass and iron coatings could come from processed lunar materials. Repeating this mounding process over many acres creates a giant sized solar collector which could easily be expanded."

An entire lunar base, including antennas, collectors, and habitat to house workers, could be about 100 kilometers in diameter. Two such bases, one located on the eastern and the other on the western face of the moon, could provide the necessary 20-gigawatt power system.

"We could use known industrial techniques of joining, building, processing, molding on the moon," said Criswell, "and the shuttle would be used only to transport necessary machinery up from Earth."

A more international moon power project is envisioned by George Mueller. He believes that solar power from the moon might become another cost-effective industry. "Although we are currently in an oil glut," Mueller said, "energy is a good reason to go into space." An alternative to a space-based solar power satellite would be a lunar-based system built of lunar materials, processed there. "Multiple power stations would gird the moon near its equator," said Mueller, "and would transmit to various receiving stations on Earth. It's more technically feasible than a power station in geosynchronous orbit because the moon is a stable platform, has lots of resources and would reduce transportation costs of materials from the Earth's surface."

Besides that, the moon's environment is ideal for solar energy generation: it is a vacuum, is always dry, and has vast areas upon which giant collectors can be placed.

Mueller believed that such solar stations could be built at "reasonable expense" with immediate prospects of payback. "It would create new wealth and help us manage our terrestrial resources better," he said. "It is a politically viable project to expand into space." International stability would be enhanced, since "power could be distributed through each nation's grid and it would be best to use worldwide transmission systems."

Human presence on the moon would cause "a lunar renaissance," thought James Burke of the Jet Propulsion Laboratory. "Specific lunar living skills have to be developed along with the technology — like special enthusiasm, leadership, self-reliance, and resourcefulness." Astronomers would be delighted with the moon's excellent observatorial possibilities while space explorers would find easy access to the further exploration of space.

Habitats might be constructed deep under the surface of the moon, according to one Air Force Academy design, or might use thrown-away shuttle propellant tanks and nuclear power. Whatever the housing might be, "crews might not want to inhabit the moon permanently at first," warned B. J. Bluth, a psychologist from California State University. They might find themselves coming down with symptoms Antarctic residents sometimes have: carelessness, fatigue, boredom, hostility, and depression. "Communication between the Earth and the moon will probably be better than it is between the Antarctic and other places on Earth," Bluth said. "And our experiences in the Antarctic may help pave the way to settling on the moon. We might begin to identify steps that can be taken to enhance the quality of life by improving the kinds of conditions that are conducive to productivity and satisfaction."

A recent study concluded that condominium-like living, ample gardens, a large lake, and a climate like that of southern California would all be desirable features of a lunar city.

Headed by Count Rinaldo Petrini, a lunar colony design workshop was held in Los Angeles in 1982. It envisioned primitive lunar base camps giving way to domed cities, once mining and other moon businesses get started. The domes would be inflatable structures made of spun fiberglass or Kevlar covered over with layers of the moon rock for protection against solar and cosmic radiation. Sunlight would be directed inside the habitat by a series of reflecting mirrors.

The domes would be connected by a network of underground tunnels where moon residents could huddle in case of a large solar storm. Pressurized workshops and unpressurized hangars would also be connected into the network of tunnels.

Residents would live in small apartments, only 18 feet wide and 22 feet long, where they can plug into Earth TV broadcasts and telephone communications and get a sense of privacy. They would even have tiny kitchens to prepare individualized meals.

Public areas would include amphitheaters, gyms, and gardens. The lake would provide the habitat's drinking and bathing water and could be used for swimming and diving as well. Small garden plots might be allocated to colony members who would enjoy cultivating food in their free time. If the domes are kept at a constant 72-degree temperature, the climate might allow for a large array of flora, including aromatic and fruiting trees, shrubs, and grasses.

"We have to see if we can adapt technology to man rather than man to technology," said Petrini.

Lung damage from lunar dust was an early concern of the space program, but according to Dr. J. H. Chalmers and Dr. E. C. Lawrence of Baylor College of Medicine, lunar dust had "overall little effect" on lung function as opposed to common Earth environmental substances like asbestos, which are much more dangerous. "The dust in west Texas is worse than lunar dust," said Dr. Lawrence," but we need to study it further."

Water is crucial to all terrestrial life, and its discovery on the moon would have profound implications for eventual human lunar settlement.

Unfortunately, it is almost impossible to submit pure science projects, whose aim is scientific knowledge, to rigid cost-effective schemes. "It is very difficult to design a valid cost-benefit test when you are dealing with the unknown," said Mueller. "There was no need for a telephone before it was invented, just as there was no need for a printing press before it was invented. Yet these two 'needless' developments proved to have an impact on society far beyond their inventor's dreams. Throughout history, we have seen the incalculable benefits from bold explorations and developments that could never have passed a cost-benefit test based upon the conventional frame of reference of the time."

Future moon programs might have to come about as a result of a coalition of space activist groups, industry, and private space entrepreneurs who can exert political pressure on apathetic government officials or gather large sums of money to finance private moon exploration. "The space groups did succeed in blocking U.S. ratification of the Moon Treaty, which would have effectively banned any kind of free enterprise on the moon," said Nathan Goldman, a University of Texas political science professor and space law expert. "Similarly, they succeeded in keeping the Jupiter-bound Galileo probe in the American budget."

"The combination of an inexorably growing population, an increased standard of living, and a consequent depletion of nonrenewable Earth resources would certainly dictate SOME consideration of space industrialization and of human expansion into the solar system," said Mueller.

NASA's former deputy administrator, Hans Mark, believes that we will go back to the moon, but only when some "enabling technology" makes it cheap and easy to get there.

Many years ago, Mark discussed mankind's return to the moon with the late Wernher von Braun. Von Braun compared the race to the moon to the race to the South Pole, in which national prestige was a prime motivator. "Twenty years after the race to the South Pole was won by Roald Amundsen of Norway," said von Braun, "nothing much happened. Scientific reasons were not enough to compel people to return there." In 1929, Commander Byrd flew to the Antarctic; the airplane, though designed for other reasons, proved to be the enabling technology that led to the eventual colonization of the Antarctic. Currently there are more than 1,000 year-round visitors to Antarctica, and about 10,000 during the summer. There are some comfortable villages representing many different nations, and there are small mining projects geared to using the Antarctic resources.

Similarly, said Mark, expendable rockets were an "expensive way to make a first trip to the moon." However, if one follows the timeline of Antarctic explorations, we can expect "the first permanent bases on the moon in 1995 and a permanent population of 1,000 on the moon by 2040." The enabling technology that will make moon settlement possible will be "a true spaceship, one that is designed solely for the purpose of flying in space," not a "hybrid vehicle" like the space shuttle, which is designed to carry people and cargo from the surface of the Earth into orbit and back.

Von Braun foresaw it when he said: "You start out for reasons of national prestige, you then start doing scientific research slowly with this new enabling technology; when you get better at it you have an international collaboration to do science, part of which is to assay the resources."

Who the moon miners and builders of lunar colonies might be is anybody's guess. Undoubtedly, they are already born and being educated. What specializations, what career preparations they should be making are matters of sheer speculation at this point. However, many people feel at NASA that the settlement of space will not be haphazard like the settlement of the American West. It will require a logical combination of economic motivation, technological know-how, and political incentive to accomplish it. At least, the economics and technology argue for a return to the moon within the next 20 years, for the purpose of permanently settling and mining it.

It is possible that among the next 1,000 people in space will be the first crew to Mars. Such a trip would require bold incentive, daring determination. A trip to Mars may be the natural outgrowth of our expansion to the moon, or it may be, like the Apollo program, begotten out of a sense of national pride. It may be, as NASA Administrator James Beggs speculated, an "international effort, in which the free countries of the world would make a contribution and share in the benefits from such a voyage."

The round trip to Mars may take two years — six months each way and a year to explore with standard chemical propellants; or it could take only a few months — two weeks out, two weeks back, and 40 days to stay with some advanced propulsion system now only on a drawing board. Supplies could be carried aboard ship or sent out ahead in vessels powered by solar sails.

There will be many hazards the Mars voyagers will have to face.

If the trip is a long one, supplies will be a critical problem. Medicines and food grow stale over time, and human beings use a great deal of water and air to stay alive. "Life support includes everything," said Phillip Quattrone of NASA's Ames Research Center, a top scientist in the life-support area. "Air revitalization, water reclamation, recycling of wastes, food cultivation, and even such necessities as protective clothing are all part of life support. We'll need carbon dioxide control, oxygen and nitrogen generation, humidity, temperature and contamination control—and that's just the beginning."

The key danger to the health of the human crew on the surface of Mars may be radiation. Even with solar shelters to protect them from severe solar storms, the astronauts could still be exposed to high doses of radiation, raising the risk of eventually getting leukemia or some other radiation-induced cancer to two or three percent. More than usual care needs to be exerted in putting together medical kits, exercise equipment, and vitamin supplements, including facilities for emergency treatment.

But there isn't an astronaut in the astronaut corps who wouldn't go tomorrow if he or she had the chance.

And it wouldn't just be the adventure. The work they would be doing on the surface of Mars would be nothing less than scouting the planet out for a new home for mankind, the most important work of this millenium if it happens to fall before the year 2000 or the next if it falls after.

With current scenarios, the spaceship would approach Mars, aiming for a narrow corridor, grazing the Martian atmosphere and using its drag to decelerate the spacecraft, a technique called "aerobraking" that does not require rocket engines. Once in orbit around Mars, the astronauts would board smaller landing craft to go to the surface of Mars, to set down softly in a cloud of red dust. The crew might remain on the surface for days, weeks, or even months, scouting the terrain on foot, in rovers or possibly in a "Mars airplane," a craft specially designed for the thin Martian atmosphere.

"In terms of its resource, Mars is more hospitable than the moon and even some places on Earth," said geochemist Benton Clark, who made a study of Viking results. Martian soil includes 20 percent silicon, 12 percent iron, three percent sulfur, and some nitrogen, carbon compounds, and minerals, containing aluminum, magnesium, titanium, and chlorine. "If there isn't life on Mars, it's not because of the elements," suggested Clark. "If man gets to Mars, he'll be able to use the raw materials there to manufacture the necessities of life: habitats, fuels, even food."

Astronauts on subsequent journeys to Mars will establish above-ground habitats by stacking cinderblocks made with Martian iron or sulfur sealed with a sulfur glaze. If radiation forces them to live beneath the surface, they could blast shelters beneath ground by using explosives made from local materials such as ammonium nitrate.

According to Viking results, hydrogen, an element crucial to human life, is very scarce on Mars. Prospectors may have to seek a source of hydrogen on Mars, or it will have to be imported in the form of hydrogen peroxide. If water is found on Mars, perhaps close to the poles, it would be possible to split hydrogen atoms from the water. Phosphorus, another valuable necessity of plant life, might be present in igneous rocks on Mars, another aim of prospecting.

Once the humble building blocks of life are found or transported to Mars, forecasters foresee rapid expansion and development of the planet. University of Texas zoologist Bassett Maguire, who has studied closed ecological systems, noted that a truly closed ecosystem could cause plant suffocation. Plants need adequate supplies of carbon dioxide just as human beings need oxygen. Animals, he believes, will be introduced to balance atmospheric gases. Rabbits would be ideal because they would fill several other ecological niches — they eat parts of plants that are indigestible to humans, their wastes can serve as fertilizer, and they can be consumed by humans as a source of protein.

Several crops would do well on Mars, such as soybean, sugar beets, rice, lettuce, onion, radishes, potatoes, yams, and various herbs. All these plants have abundant yields, don't require much space to grow, and are high in protein and vitamins. Genetically engineered superhybrids, unknown today, may become staple crops on Mars, as the ecosystems are refined and modified. In the past it was thought wise to eliminate bugs, fungi, and all forms of bacteria from plants grown in space, but recent wisdom dictates that these tiny life forms help clear away dying roots, leaves, and stems, preventing waste buildup in plants. Such space testing of agriculture may yield crops for hostile, arid growing areas on Earth as well.

There is only one thing that prevents the United States from investing in the exploration of Mars: money. Although it wouldn't cost as much as the Apollo program — because the research and development money required would be greatly reduced by the shuttle transportation system and the space station technology, it would cost at least 20 to 40 billion dollars. As NASA Administrator James Beggs pointed out, the trip to Mars may be undertaken by the United States and its allies, in which all the free nations of Earth could share in the pride of landing people on Mars. People who have never stopped dreaming of going to Mars point out that there are good reasons to go there: to wrest more secrets from it, to use its resources to enrich all humankind, to escape the specter of eventual resource exhaustion on our own planet, to expand our knowledge of the universe, and to open the gateway to the metal-rich asteroids beyond Mars and the moons of Jupiter and Saturn beyond them. A few see their children mining Mars, their grand-children settling it, their great-grandchildren forming a sovereign government there. What we learn from the exploration, the settle-ment, the civilization, and ultimately the metamorphosis of Mars could change human beings themselves beyond the ways in which established space stations and moon colonization might. "Space is the cathedral-building of our age," said Dr. B. J. Bluth, a psychologist with a specialty in space habitability. "Even if the chances of survival on Mars were less than calculated, isn't it worth a life to seed the universe with intelligence, to have all humanity grow as a result of an individual's contribution?"

One exciting reflection came from former astronaut and U.S. Senator Harrison Schmitt. "Very probably," said that tireless cham-pion of the space program, "the parents of the first native-born Martians are alive today and living among us."

It is certain that the first Mars voyagers are already or will soon become members of NASA's astronaut office who do not yet have the vision to see, as through a glass darkly, that they will be the Chosen.

13

How Do I Get To Fly In Space?

"How do I become an astronaut?" is the question most often asked by young visitors at the Johnson Space Center. To begin with, persons must apply when NASA issues a call for astronaut applications. The astronaut corps is likely to maintain an active list of about 80 astronauts at any one time. Additional needs might arise with, say, additional Orbiters coming on-line. Currently the *Columbia*, the *Challenger*, and the *Discovery* are operational, and the *Atlantis* will be by late 1985. When all four Orbiters are operational, launches are expected to be very frequent — once a month and in a few instances, twice a month. Additional Orbiters could certainly enlarge the necessary number of working astronauts to more than a hundred. Crews can range from the usual four up to eight for Spacelab flights. Astronauts could expect to fly about two or possibly even three times a year, depending on their specialty.

NASA's call for astronauts has never been at predictable intervals in the past. The last, made in 1983, resulted in only 17 new astronaut candidates. Attrition from the astronaut corps as a result of retirement, military transfer, or personal choice does happen periodically, but it happens so sporadically that NASA is considering maintaining a regular file of applicants who will be screened and hired as the need warrants. So far, however, that is only an idea under advisement. From the standpoint of screening and training, it is easier to take a group of astonauts than individuals one at a time.

In any case, NASA has certain restrictions for astonaut applicants that are unlikely to change in the near future.

Age restrictions call for people under 40 and over 18. People under 18 are unlikely to have the education and the experience for the job. Because astronaut training takes a few years and because human eyesight and hearing begin to degrade from about 50 years old on, people over 40 are considered too old to give NASA more than a few years of useful service. Education is important. A bachelor's degree is a minimum requirement. But of the 35 astronauts selected in 1978, 10 had Ph.D.s in engineering or science. Three had M.D.s, 15 had an M.A. or M.S., usually in an engineering specialty. Only seven had the minimum B.S. degree, of whom all were of the pilot category.

Although there are no space specialties as such in universities that would guarantee a job in space, the physical and engineering sciences far outstrip the other fields of endeavor. People from the behavioral sciences, social sciences, or the humanities are, and probably always will be, considered unsuited for astronaut duty.

Aviation background figured heavily into the current astronauts' backgrounds. In the pilot category all had either fighter pilot or test pilot experience, or both. Many were top of their class at test pilot school. Many of the scientist-type mission specialists also had private flying licenses. NASA has never chosen pilots from commercial aviation, although the shuttle astronaut job is similar to that of an airline pilot in many respects. An organization of pilots that believes that commercial pilots might be a source of future shuttle pilots is fighting this. Women, who are still not allowed into test pilot or combat positions, are developing ways to try to sidestep this dictum. Although NASA may eventually agree to hiring airline pilots and women pilots, most insiders feel that this is unlikely soon.

The rigid medical tests are the great disqualifiers. They include extensive electrocardiographic studies during hyperventilation, carotid massage, breath holding, running, and so on; x-rays of upper gastrointestinal tract, spine, chest, and skull; pulmonary breath tests; complete blood workup and urinalysis; detailed examination of sinuses, larynx, and eustachian tubes; vestibular studies with rotating chairs and so on; neurological examination; dental exam; proctosigmoidoscopy for men; surgical evaluation; electroencephalographic studies; visual and hearing tests; and psychological interviews.

In the 1978 selection, 28 finalists were crossed off for inadequate vision; 13 for cardiovascular problems; four for endocrine problems of the prediabetic, hypoglycemic kind; four for previous surgery; three for neurological disorder; and a few in all the other areas of concern except dental. In the 1980 selection, vision and cardiovascular problems were the leaders in disqualifying medical problems, and a large number, 12, of finalists were disqualified for "psychological reasons," either claustrophobia or other latent problems. Some were disqualified for attitude problems. One unfortunate candidate wanted to dicker for a higher salary, showing, perhaps, lack of motivation and dedication. During the selection for the 1984 candidates, people who "rested on their laurels" were shuffled out early in the process. "Self-starters" and "go-getters" are the ones who get in.

It is unlikely that people with chronic disorders like diabetes, epilepsy, or hypoglycemia will ever be allowed into the astronaut corps. The types of cardiovascular problems that eliminate people are often too subtle to pick up on ordinary physical exams and make themselves identifiable only with extraordinary cardiovascular workups. People with heart trouble will probably not make it into space.

On the assumption that an applicant can pass the medical and psychological tests and has the educational and professional background, he or she must still be interviewed by a board made up of the director of flight operations, various astronauts, physicians, and other personnel. They score on academics, pilot performance, health, character, and motivation. Motivation is very important to all on the board, and character is judged by several kinds of questions. People who are well-balanced, have a wide variety of interests and hobbies, and are aggressive and active participants of life have a better chance than the reclusive, the passive, and the supine.

Rejection from one astronaut selection does not mean that a person can never become an astronaut. In fact, half of the astronauts selected in 1980 had been rejected in 1978 only for lack of job slots. Many of those worked to improve their qualifications between selections. Bill Fisher, for instance, bettered his chances by adding a master's of engineering to his M.D. degree. Persistence does pay — he was selected.

And even if one does not make it into the astronaut corps, one can still find other ways to get into space. Many people who were rejected by NASA — Byron Lichtenberg, Mary Helen Johnston, Millie Wylie Fulford, and Charles Walker — made it into space as payload specialists.

Following are remarks I have gathered from people within the space program on the future pathways into space:

James Beggs, NASA administrator: "I think there will be regionalisms in space. The commercial community may differ from the scientific community up there. The astronaut corps has shifted a great deal over the years. The commercial space groups will be different, and as we learn to work in space long-term, many people can fly back and forth. I don't see a melting pot on the new frontier. This new frontier is going to be restricted to people who are motivated, who have discipline to work in this environment. Space will develop virtues and encourage certain virtues: discipline, dedication, high achievement. Perhaps future space people will not be quite the high achievers they are now, although it will still take discipline and motivation. People who reach for this objective will still have to be over the norm. No matter how routine space becomes, it'll be a relatively challenging task requiring sophisticated technological skills. In the next twenty years, I really don't think the average man will inhabit space. Maybe later when you have large colonies in space, once you get a computer that thinks, that monitors systems — then a large colony will be possible, not before. By the year 2000, I see not yet a city, but an international village working together."

Larry Bell, space architect, University of Houston: "I see people from all walks of life working in space in the year 2000: doctors and nurses, housekeepers and galley workers, maintenance crews, scientists and researchers, technicians of all types, various kinds of space pilots, satellite assembly workers, communications operators."

Pat Musick Carr, professor of human sciences, University of Houston: "People who adapt. I think the best problem solvers are balanced people — both convergent and divergent thinkers, since both kinds of thinking are involved in problem solving. Some personal liberty will be permitted, but there should be strong leadership and plenty of creative kinds of activities available as a safety valve."

George Abbey, head of flight operations, JSC: "The astronaut program is probably the shortest route to get into space. The payload specialist may invest a lot of time in a very specialized field and never go. Obviously they're going to need a good background in engineering and the physical sciences because the people they're competing with are talented in those areas. You have a lot of people with excellent backgrounds and if you're going to compete, you have to know these subjects very well. And the other is good experience, someone who's actually gone out and applied that knowledge to real situations. The kind of individual who wants to be an astronaut has to be competitive and will have to be very dedicated. They have to put up with the sacrifices. They might have to sacrifice a huge financial reward. There are people who come to the program who could be doing much better financially. And from the standpoint of having time off or having lighter work loads — it takes time and effort and a good commitment on the part of the individual, because the greater part of what astronauts do is not enjoyable. Flying is glamorous, but most hours, not visible to the public, is just plain tedious work."

Alan Bean, artist, former astronaut: "Your prime interest in life has to be doing space. If it isn't, you'll get discouraged. You've got to be a generalist, but one that has the capability of dealing with highly technical work. You must be able to work as a member of a team. It helps to be organized in the way you do things, consistent in performance, careful and professional in the way you do things. It's okay to be competitive but not uncooperative."

Bonnie Dunbar, mission specialist astronaut: "You have to have humor and patience."

Ron Grabe, pilot astronaut: "Your thinking has to be geared toward flying, and you always have to think ahead of the plane. Test pilots are exposed to making decisions fast, and it doesn't hurt to have adaptability in flying, that is, experience in all kinds of aircraft."

Vance Brand, pilot astronaut: "Bright, hard-working. A self-starter with lots of determination. It's important to do well in school. It's good to show an early interest in space by gearing thesis papers and majors in space sciences. Take the fork in the road, whether in education or experience, which leads to space."

Rick Chappell, chief mission scientist for Spacelab 1: "If Spacelab is utilized, payload specialists may come from biomedical sciences, plasma physics, astrophysics, materials sciences, fluid dynamics, Earth observations, and atmospheric sciences. The potential payload specialist would have to have a doctorate in that area and could work on space-related projects. I'd love to see some space career route develop but it depends on the budget."

Byron Lichtenberg, payload specialist, Spacelab 1: "Any field that has science applications in space has the potential for a payload specialist. A person interested in flying in space would need to write a proposal, get it accepted, and make a good case for why NASA should fly the experiment in space. When we get a space station, it'll be much easier to do this — maybe just a matter of putting an experiment in a suitcase and taking a shuttle. A real avalanche of scientific ideas is possible with a space station."

James Rose, project manager for McDonnell-Douglas' pharmaceutical project, EOS: "A commercial payload specialist will be someone highly experienced in the operation of the equipment."

Mary Helen Johnston, backup payload specialist for Spacelab 3: "You're going to have to be a specialist to be a payload specialist. My advice is to pick an area that looks like it's going to have a lot of spaceflights, say, biomedical, pharmaceutical, and so on, and get involved in it. Someday, special technical companies may be a source of space employment, recruiting-type companies who can go to NASA and say 'We have a group of people who can fly materials processing stuff or medical experiments.' But that's still a long way off. In the meantime, you have to subconsciously aim yourself in the direction you want to go. I did and it was just fortuitous that I happened to be there when the opportunity came up."

Taylor Wang, payload specialist for Spacelab 3: "To try to become a payload specialist depends a lot on accidental circumstances. Being a noncareer astronaut mostly involves being in the right place at the right time, where there's an experiment that a career astronaut does not have a background in. You aim to do scientific experiments, but you can't say 'I'm planning to go into space,' because that opportunity may never present itself."

Physician-astronaut William Thornton: "I would not exclude anyone who I thought was physically and mentally qualified, and I would not put an age limit on it. Good physical condition matters more than chronological age."

Program manager for the Air Force Manned Spaceflight Engineer Program: "Get a good technical education. Perform to the best of your ability. Seek out challenging, space-related assignments. Pursue a graduate education. Apply for each selection opportunity. The current MSEs are a special breed. They are highly intelligent, aggressive, ambitious, hard-working overachievers."

Soviet cosmonauts Vladimir Lyakhov and Aleksandr Aleksandrov, veterans of long-term spaceflight: "Kindness and professionalism are the qualities I hold most important."

Guion Bluford, first black American astronaut: "The path to space will mostly be through the Astronaut Office in the next several years. Opportunities are limited, so it's better to dedicate yourself to a career that you want, and if you go into space, fine. Competition is very keen. There are lots of really good people out there. It's good to be a goal-oriented, very persistent type of person, who can set goals a few years down the pike and pretty much achieve them."

Ellison Onizuka, the first Asian-American astronaut: "Awareness of space is reaching out to all levels of society, especially the schools. The decision to try to become an astronaut has to be made early, no later than high school, because you'll be competing with people who made that commitment very early. As we progress through the years, we're going to find that astronaut classes will be composed one hundred percent of people who chose to try for the astronaut corps very early. There won't be people in it like now, who got the job without devoting their education to it."

Franklin Chang, first Hispanic astronaut: "When I talk to students, I tell them that all kinds of people will be going into space. The pioneers are going to be the technical people who will get everything set up. But after that, space stations will become bigger and bigger, and more people will come up. I bet in 20 or 30 years, there will be all kinds of people in space, and that makes a lot of sense because space shouldn't be limited to a certain group of people. We'd be discriminating again and that's not right. We should make space accessible to everybody. In 40 years, the frontier won't be in near-Earth orbit; it will be out somewhere in the solar system."

If you take this advice to heart, you can ask for an application to:
Astronaut Recruiting
Mail Code AP-4
NASA — Johnson Space Center
Houston, TX 77058
Let me know if you make it. I'd like an exclusive interview.

Epilogue

People who work for the space program are a breed apart from people in other walks of life. Very few of them are motivated by money. Even fewer are motivated by power. Their dogged dedication to the future is something they only half-comprehend. If you ask them — the aerospace engineers, the astronaut trainers, the equipment builders — why they do what they do, you might get half-mumbled responses about the excitement of being involved in the space program, of liking the work, or a shrug and an "I dunno" response. When the old, hackneyed criticisms are leveled at them about the high cost of the space program, there is sometimes a dazed look in response, as if they cannot comprehend how anybody can think badly of the space program.

It is because they live in the future. More than any other sector of American society, the people in the space program relate to the future as something tangible. For a spacecraft designer, sitting in an office at the Johnson Space Center, the line drawing of an orbital transfer vehicle, a kind of ferry-ship that will haul satellites to and from space stations and out into geosynchronous orbit, is real. Never mind that Congress has not funded even the space station that the ferry will serve. Never mind that the Office of Management and Budget has taken yet another swipe at the NASA budget. Never mind that the atomic bomb might fall, that the national debt is at a new level, that the dollar is under attack on foreign markets, that the youth of America, according to SAT scores, are dumber than ever. Never mind that she drives a rusting compact car whose gears slip, or that she needs caps on her teeth. For her, the future is now, the precise lines of the little ferry-ship that won't fly for another decade need to be designed now — its dimensions, its propulsion system, its life support system, its grapplers and remote manipulating arms, and its safety features to protect its pilot. The future is their present. Will this craft be used only to haul satellites around or will it be involved in mining expeditions to the moon as well? What materials will be available for micrometeoroid shielding? These are the essential questions.

The same is true of the physiologist studying decalcification of the bones in bedrest studies at Ames Research Center, the planetary scientist looking at the moons of Jupiter at the Jet Propulsion Laboratory, the engineer analyzing telemetry at Goddard, the equipment specialist running ground tests at Marshall Spaceflight Center, the mission controller writing the spacecraft operating manuals and watching inflight performance at the Johnson Space Center, and even the bureaucrat at NASA Headquarters in Washington, D.C., trying to answer the doubts, fend off the budgetary attacks, and sell the space program to all who still, after all these years, do not believe in it. For them the excitement of the space program did not end when Alan Shepard was launched into space or when Neil Armstrong and Buzz Aldrin stepped out onto the surface of the moon. It is not a 30-second spot on the nightly news once in a while or a George Lucas film every three years. The excitement is now, day in and day out. Excitement for them is laying the foundation for the future: the procedures astronauts will use for the next three decades in spacewalking, space construction, and satellite repair; the equipment that scientists and manufacturers will use on a space station in the 1990s; and the design of spacecraft that will get us around Earth's neighborhood in the next century. Most have not ever stopped dreaming of the day we'll go to Mars.

This book is only about a small portion of the space program, the people who will fly into space. It is about the choices they made, the selection procedures that will get them there, the work they will be doing. They are the future, their work is the future. They will fly into space because they made their own luck. They never stopped beating on future's door, never gave up the dream. But they are lucky, too, because their opportunities are predicated on the efforts of all those nameless people who drive rusty cars and will never fly in space.

I believe that being a person who works on the space program but never actually flies in space is as dignified, as important, as essential, and even as exciting as being one who does fly. And now more than ever, there are more opportunities to get involved in space work.

The Air Force employs a large number of people from different walks of life to carry out its interests in space. When I was in Cheyenne Mountain (at NORAD — the North American Defense Command) looking into Air Force space careers for *Science Digest*, I talked to people involved in satellite tracking who had degrees in English, history, and physical education. The Air Force had trained them to be managers and specialists in satellite tracking. Later, among the manned spaceflight engineers, I found Air Force officers with degrees in business administration and health, along with the engineers and scientists. Undoubtedly, intelligence efforts could use expert radio people, multilingual people, and computer wizards with a flare for linguistics.

NASA employs a wide range of people in its various centers around the country — not only people who work on the nuts-and-bolts aspects of spaceflight, but also artists to do the drawings of future spacecraft, public relations people eager to give out information, psychologists and doctors looking at habitability issues for future missions, and many more. NASA has a certain number of contracts it lets every year, having to do with wide-ranging subjects — from the development of future technologies to the development of the human spirit's creative potential in space.

There are few areas of science that will avoid involvement in space. Already universities the world over are proposing space experiments from all areas of science, and people involved in these always have their eyes on the future.

Private corporations, for the first time ever, have the administration's blessing and encouragement to get involved in space. Private companies with their own space program, like Space Services Incorporated of Houston, recapture the ground-floor excitement of the early days of space travel. Entrepreneurs who analyze the space market and spot a clear need for a certain item can found new companies — whether the item is a modest, middle-sized rocket or a modest space station. Opportunities for manufacturing new products probably require a large amount of venture capital, but workers will have to build the equipment, and test, sell, and market the product.

Of course, there are a large number of contractors employed by NASA to provide the various necessities of the space program — both goods and services. These range from large aerospace companies like Rockwell, McDonnell-Douglas, Boeing, and Lockheed to much smaller ones. These companies employ not only young engineers and managers but also tour guides, little old ladies who sew well (parts of spacesuits are hand-sewn), and artists who illustrate proposed future space equipment. Now that the way has been cleared to develop space industry, some of these companies may look to other space ventures of their own.

I'm sure thousands of other opportunities for spacework I haven't even thought of exist both here and abroad. It is silly to think that if one cannot fly in space, one cannot be deeply and intimately involved with the space program. Americans young and old should never think that the space program is something they can't be involved in for a long time. Politicians enamored of the space program could certainly do their share in voting for space; citizens can support these space-loving politicians, can join space activist groups all over the country, and can even write ther Congressional representatives; investment bankers can think of how best to invest their clients' money in space ventures; senior citizens and handicapped people should push space for the medical benefits it will bring; labor unions should push it for the jobs it will open; housewives and educators should rally to it for the expansion of knowledge and educational opportunities it will provide to our children and our children's children; minorities should support the new opportunities it will offer and should insist on being a large and influential voice in the development of the future; poor nations should welcome it for the new riches it will yield, for the natural resource development and the communications revolution that will fight famine and ignorance for their people; people the world over should herald it as the one means we have to develop as a species and the only means we have to preserve the natural resources and avert climatic disaster.

Why not you? Why not now?

I think this is one of the few times in the history of mankind that common people can have an impact on the future, when they'll really matter. Individuals fortunate enough to be among the next thousand people in space will open the pathway to space for others. The Earth-bound will support their efforts economically, politically, and culturally until they too can follow. As early settlers, all can have an impact on the future by exploring its new environment, reaching out for its new opportunities, and even founding the intellectual history of space.

The space environment will have its own lessons to teach. Certain human virtues will be affirmed, certain human traits will be discredited, certain human perspectives will be expanded. One astronaut once told me that the view from space makes nations seem provincial, that the global perspective is something he could apprehend in space. I believe that the great hope of humankind — that we can all live in peace on our planet — is something that may be achievable only in a space civilization.

Whatever lessons we learn and however we develop as a species, the present is less like a lesson and more like an unconscious wave. The next thousand people in space will be like the exhilarating foam pushed forward onto the threshold of land by that huge, deep wave whose origins are far out to sea, nobody knows just where. The wave moves relentlessly toward the shore as a whole, unconscious of the powers of sun and moon that guide its restless rhythms.

Bibliography

Aldrin, Edwin and Wayne Warga. 1973. *Return to Earth*. New York: Random House.

Allen, Joseph and Thomas O'Toole. 1983. "Joe's Odyssey." *OMNI* (June), pp. 60-63, 114-116.

Collins, Michael. 1974. *Carrying the Fire*. New York: Farrar, Straus, and Giroux.

Cooper, Henry. 1976. *A House in Space*. New York: Holt, Reinhart, and Winston.

Hevener, Phil. 1977. "A Woman's Place in Space." *Newsday* (October 17), p. 4.

National Aeronautical Institute. *Who's Who in Aviation and Aerospace*. 1983. Boston: Warren, Gorham & Lamont.

Navias, Rob and Patricia Walker. 1983. "A Woman with the Right Stuff." Quoted in the *Congressional Record* E 2611-12 (June 1), pp. 12-13.

Nicogossian, Arnauld and James Parker. 1982. *Space Physiology and Medicine*. Washington, D.C.: NASA.

Oberg, James. 1981. *Red Star in Orbit*. New York: Random House.

O'Toole, Thomas. 1978. "Thirty-Five New Guys." *Washington Post* (April 6), pp. 4-8.

Rice, Berkeley. 1983. "Space-Lab Encounters." *"Psychology Today* (June), pp. 50-58.

Santy, Patricia. 1983. "The Journey Out and In: Psychiatry and Space Exploration." *American Journal of Psychiatry* (May) 140(5): 519-527.

Turnill, Reginald, Editor. 1984. *Jane's Spaceflight Directory*. London: Jane's Publishing Company.

Welch, Brian. 1983. "Robert T. McCall." *JSC ROUNDUP* (October 14), p. 3.

Scientific Conferences Pertaining to This Book

A Case For Mars I, Boulder, Colorado. 1981.
Aerospace Medical Association Convention, Houston, 1983.

AIAA Business Symposium, Johnson Space Center, Texas. 1983.
American Association for the Advancement of Science. Houston, 1979.
Lunar and Planetary Science Conference, Houston, 1983.

Press Conferences

All pre-flight, in-flight and post-flight conferences for (with names of crew):
STS-1 (Young and Crippen)
STS-2 (Engle and Truly)
STS-3 (Lousma and Fullerton)
STS-4 (Mattingly and Hartzfield)
STS-5 (Brand, Overmyer, Allen, and Lenoir)
STS-6 (Weitz, Bobko, Musgrave, and Peterson)
STS-7 (Crippen, Hauck, Ride, Fabian, and Thagard)
STS-8 (Truly, Brandenstein, Gardner, Bluford, and Thornton)
Spacelab I (Young, Shaw, Garriott, Parker, Lichtenberg, and Merbold)
STS-41 B (Brand, Gibson, Stewart, McNair, and McCandless)
STS-41 C (Crippen, Scobee, van Hoften, Hart, and Nelson)

Phone or Personal Interviews With:

Abbey, George (Director, Flight Operations, Johnson Space Center)
Allen, Joseph (astronaut)
Bean, Alan (astronaut and artist)
Beggs, James (NASA Administrator)
Bluford, Guion (astronaut)
Bluth, B.J. (sociologist with specialty in space)
Brand, Vance (astronaut)
Carr, Gerald (astronaut)
Carr, Pat Musick (psychologist and artist)
Chang, Franklin (astronaut)
Chappell, C. Richard (chief scientist for Spacelab I)
Cline, Cindy (liaison with West German space program, NASA Headquarters)
Covault, Craig (aviation and space journalist)
Cowles, Joe (biologist)
Criswell, David (consultant with specialty in lunar development)
Dalton, Maynard (spacecraft designer)
Degioanni, Joseph (NASA flight surgeon)
Dooling, Dave (space and science journalist)

Dunbar, Bonnie (astronaut)
Dunbar, Mrs. (astronaut's mother)
Fink, Daniel (member NASA Advisory Council)
Fries, Sylvia (historian, member NASA Advisory Council)
Gazenko, Oleg (Head of Space Medicine, USSR)
Gibson, Robert (astronaut)
Goleby, Ron (liaison with Japanese space program, NASA Headquarters)
Grabe, Ron (astronaut)
Hoffman, Jeffrey (astronaut)
Hughes, Frank (Head of Astronaut Training)
Jackson, Jake (human factors, Johnson Space Center)
Johnston, Mary Helen (payload specialist for Spacelab 3)
Kerwin, Joe (astronaut)
Lichtenberg, Byron (payload specialist for Spacelab I)
Lillibeck, Norman (metallurgist, John Deere Co.)
McClure, Bill (Astronaut Scheduling)
McNair, Ron (astronaut)
Michener, James (author, member NASA Advisory Council)
Naugle, John (NASA chief scientist, member NASA Advisory Council)
O'Donnell, William (NASA Public Affairs, Washington, D.C.)
Onizuka, Ellison (astronaut)
Peterson, Don (astronaut)
Petrini, Rinaldo (architect)
Prinz, Dianne (payload specialist for Spacelab 2)
Quattrone, Phillip (chief of life support research, Ames Research Center)
Ramsland, Russell (executive director of Microgravity Associates)
Rappole, Clinton (participant in Spacelab project)
Resnik, Marvin (astronaut's father)
Ride, Sally (astronaut)
Rose, James (Head of McDonnell-Douglas EOS project)
Seddon, Rhea (astronaut)
Sullivan, Mr. (astronaut's father)
Tanner, Treive (contract monitor on Ames space psychology studies)
Thornton, William (astronaut)
Tingle, Mrs. Riley (astronaut's mother)
Trotti, Guillermo (space architect)
Truly, Richard (astronaut, consultant for NASA Advisory Council)
Walker, Charles (industrial payload specialist for McDonnell-Douglas)
Wang, Taylor (payload specialist for Spacelab 3)
Woodard, Daniel (physician, human factors engineer)
Workman, Gary (physical chemist for John Deere Co.)

Zweig, Roger (NASA pilot)
25 other space experts both outside and inside NASA who wished for their
    remarks to remain anonymous

Attended speeches given by:

Chalmers, J.H. and C.G. Lawrence (physicians)
Chappell, Richard (chief scientist of Spacelab I)
Clark, Benton (planetary scientist)
Criswell, David (lunar specialist)
Davis, Hubert (economist, Eagle Engineering)
Goldman, Nathan (professor, specialist in space law)
Haskin, Larry (professor of geology)
Kulpa, Major General John (Air Force General)
Mark, Hans (deputy director of NASA)
Matsnyev, E.T. (Soviet space doctor)
Mueller, George (former vice president of Burroughs Corp.)
Musgrave, Story (astronaut)
Natani, Kirmach (developer of selection procedures for Air Force)
Petrini, Rinaldo (architect)
Prinz, Dianne (payload specialist for Spacelab 2)
Quattrone, Phillip (Head of life support research, Ames)
Rose, James (Head of EOS program, McDonnell-Douglas)
Rudisill, Marianne (human factors engineer, Lockheed)

Index